Information and Instructions

This shop manual contains several sections each covering a specific group of wheel type tractors. The Tab Index on the preceding page can be used to locate the section pertaining to each group of tractors. Each section contains the necessary specifications and the brief but terse procedural data needed by a mechanic when repairing a tractor on which he has had no previous actual experience.

Within each section, the material is arranged in a systematic order beginning with an index which is followed immediately by a Table of Condensed Service Specifications. These specifications include dimensions, fits, clearances and timing instructions. Next in order of arrangement is the procedures paragraphs.

In the procedures paragraphs, the order of presentation starts with the front axle system and steering and proceeding toward the rear axle. The last paragraphs are devoted to the power take-off and power lift systems. Interspersed where needed are additional tabular specifications pertaining to wear limits, torquing, etc.

HOW TO USE THE INDEX

Suppose you want to know the procedure for R&R (remove and reinstall) of the engine camshaft. Your first step is to look in the index under the main heading of ENGINE until you find the entry "Camshaft." Now read to the right where under the column covering the tractor you are repairing, you will find a number which indicates the beginning paragraph pertaining to the camshaft. To locate this wanted paragraph in the manual, turn the pages until the running index appearing on the top outside corner of each page contains the number you are seeking. In this paragraph you will find the information concerning the removal of the camshaft.

More information available at haynes.com

Phone: 805-498-6703

Haynes Group Limited
Haynes North America, Inc.

ISBN-10: 0-87288-562-3
ISBN-13: 978-0-87288-562-2

Cover art by Sean Keenan

CSH-3, 8S1, 14-24

SHOP MANUAL

COCKSHUTT

MODELS 35-40D4 GOLDEN EAGLE

Engine serial number is stamped on right side of cylinder block. Tractor serial number is stamped on left side of main frame.

★ NOTE ★

This manual provides methods for servicing the 35 and 40D4 tractors which, except for the engine and accessories covered in this manual, are serviced in the same manner as the similar models 40 and 50 covered in I&T Cockshutt Shop Manual No. CS-2 or official Cockshutt Service Manual No. SM-254.

In the index listed below, the first column lists the tractor components, the next two columns list the beginning paragraph number and the last column lists the manual number (or this manual) in which the desired paragraph containing the service information will be found.

INDEX

CONDENSED SERVICE DATA

Tractor Models	35	40D4
GENERAL		
Engine Make	Hercules	Perkins
Engine Model	GO-198	L4
Number of Cylinders	4	4
Bore—Inches	3¾	4¼
Stroke—Inches	4½	4¾
Displacement—Cubic Inches	198	269.6
Compression Ratio	6.5:1	17.5:1
Pistons Removed From	Above	Above
Main Bearings—Number of	5	3
Main and Rod Bearings Adjustable?	No	No
Cylinder Sleeves	Dry	Wet
Forward Speeds	6	6
Electrical System Voltage	6	12
Battery Terminal Grounded	Pos.	Pos.
TUNE-UP		
Firing Order	1-2-4-3	1-3-4-2
Valve Tappet Gap	0.010H	0.010H
Inlet Valve Seat Angle	45°	44°
Exhaust Valve Seat Angle	45°	44°
Ignition Distributor Make	A-L	...
Ignition Distributor Model	IAD-4043	...
Breaker Contact Gap	0.020	...
Ignition Timing—Retard	2° BTC	...
Injection Pump Make	....	C.A.V.
Injection Pump Model	...	BPE
Injection Timing	...	21° BTC
Spark Plug Make	Champion	...
Model	J6	...
Electrode Gap	0.025	...

*Normal end play of 0.021-0.038 is non-adjustable.

Tractor Models	35	40D4
TUNE-UP—(Continued)		
Carburetor Make	Zenith	...
Model	267J8	...
Float Setting	1 5/32	...
Engine No Load RPM	1850	1850
Engine Loaded RPM	1650	1650
PTO Loaded RPM	530	530
BP Loaded RPM	1000	1000
SIZES—CAPACITIES—CLEARANCES (Clearances in thousandths)		
Crankshaft Journal Diameter	2.497	2.9985
Crankpin Diameter	1.9875	2.7485
Camshaft Journal Diameter, Front (No. 1)	2.0535	2.057
Journal Diameter (No. 2)	2.0535	2.047
Journal Diameter (No. 3)	2.0535	2.037
Journal Diameter (No. 4)	2.0535	...
Piston Pin Diameter	1.1246	1.438
Valve Stem Diameter	0.373	0.374
Main Bearings, Diameter Clearance	0.9-3.0	2.5-4.8
Rod Bearings, Diameter Clearance	1.0-3.0	2.5-4.8
Piston Skirt Clearance	3.0-3.5	...
Crankshaft End Play	5.0-10.0	4.5-11.5
Camshaft End Play	2.0-7.0	*
Camshaft Bearing Clearance	1.5-3.5	4.0-7.0
Crankcase Oil—U. S. Quarts	4.8	8.4
Crankcase Oil—Imperial Quarts	4	7
Cooling System—U. S. Gallons	4.5	3.6
Cooling System—Imperial Gallons	3.75	3
Transmission and Differential—U. S. Quarts	†38.4	†38.4
Transmission and Differential—Imp. Quarts	†32	†32

†When equipped with P.T.O. add 10 Imperial pints or 12 U. S. pints.

ENGINE AND COMPONENTS

R&R ENGINE WITH CLUTCH

Model 40D4

42A. To remove the engine and clutch as a unit, first drain cooling system and if engine is to be disassembled, drain oil pan; then, remove grilles, remove front and center hood as a unit and proceed as follows: Remove adjusting cap from front of steering gear housing and withdraw cotter pin from steering shaft. Loosen the steering shaft center bearing clamp. Turn the steering worm shaft out of mesh with the sector and pull the shaft forward as far as possible. Refer to Fig. CS24. Then, using a punch, remove the pin retaining steering wheel to steering shaft, remove steering wheel and withdraw the steering worm shaft from tractor. Disconnect radiator hoses and remove radiator and fan. Disconnect wire and cables from starter solenoid, oil pressure gage line from junction on left side of cylinder block, stop control cable from injection pump, and wires from the generator regulator. Disconnect heater cable, air cleaner hose and throttle rod from venturi and unclip wiring harness from side of cylinder block. Disconnect fuel supply and return lines from fuel tank and disconnect the fuel line connecting the fuel transfer pump (lift pump) to the fuel filter. Disconnect the heat indicator bulb from rear of cylinder head. Unbolt the fuel tank support from cylinder block and block-up between fuel tank and transmission. Remove batteries, battery boxes and clutch housing cover. Disconnect the clutch rod, then support both halves of tractor and unbolt engine frame from transmission.

Fig. CS24—Removing the steering (worm) shaft in preparation for removing the radiator on models 35 and 40D4.

Roll front half of tractor forward and securely block-up under engine frame. Unbolt timing gear case cover from engine frame and be careful not to mix or lose shims which may be installed between gear case cover and frame. Remove the engine rear mounting bolts and using a drift, bump both of the engine locating dowels up and out of the engine frame. Attach hoist to engine lifting plates and remove engine from frame.

The extent to which the engine is disassembled subsequent to its removal will, to some extent, govern the installation procedure. In general, however, the engine can be installed by reversing the removal procedure.

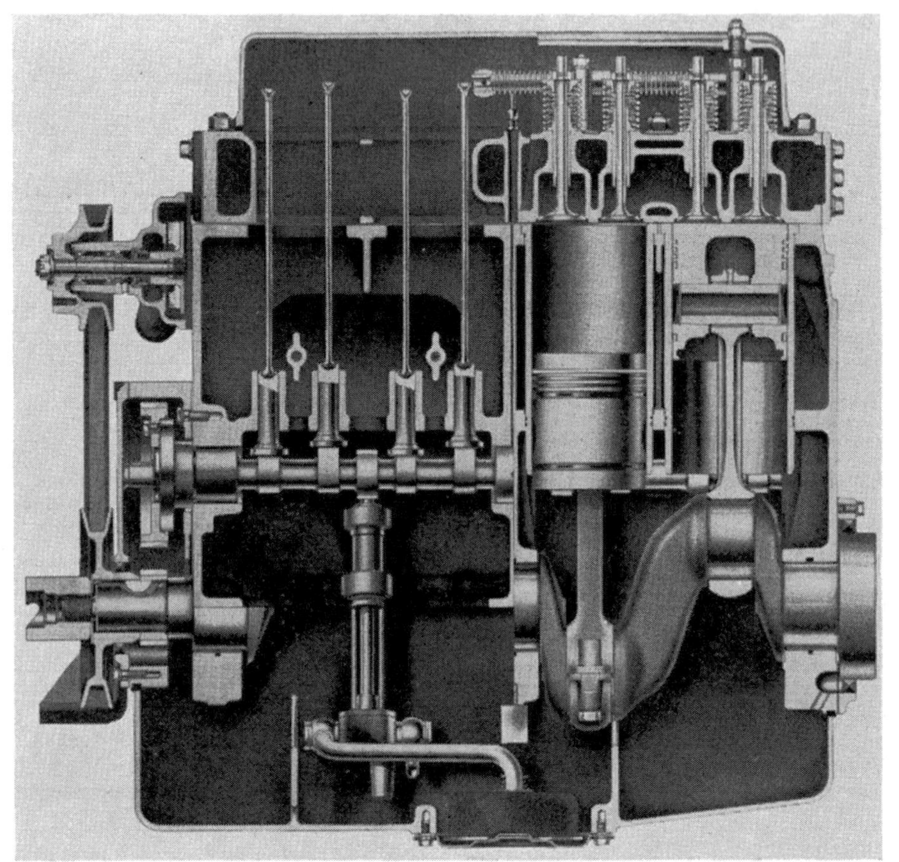

Fig. CS25—Side sectional view of L4 Perkins engine used in model 40D4 tractors. The unit is equipped with precision insert type bearings, wet sleeves and a pneumatic governor. Injection equipment is manufactured by C. A. V.

Model 35

42B. To remove the engine and clutch as a unit, first drain cooling system and if engine is to be disassembled, drain oil pan; then, remove grilles, remove front and center hood as a unit and proceed as follows: Remove adjusting cap from front of steering gear housing and withdraw cotter pin from steering shaft. Loosen the steering shaft center bearing support. Turn the steering worm shaft out of mesh with the sector and pull the shaft forward as far as possible. Refer to Fig. CS24. Then, using a punch, remove the pin retaining steering wheel to steering shaft, remove steering wheel and withdraw the steering worm shaft from tractor. Disconnect radiator hoses and remove radiator and fan. Disconnect wiring harness from starter, coil and generator regulator. Shut off the fuel and remove fuel line. Disconnect the throttle rod from governor arm and air cleaner hose and choke rod from carburetor. Disconnect the oil pressure gage line from right side of cylinder block and heat indicator bulb from cylinder head. Unbolt the fuel tank support from cylinder block and block up between fuel tank and transmission. Remove the clutch housing top cover and the dust shield from under the engine frame.

Remove the engine front mounting bolts and be careful not to mix or lose shims which may be installed between gear case cover and frame. Remove the engine rear mounting bolts and using a drift, bump both of the engine locating dowels up and out of the engine frame. Attach hoist to engine in a suitable manner, slide engine forward and lift same from tractor frame.

The extent to which the engine is disassembled subsequent to its removal will, to some extent, govern the installation procedure. In general, however, the engine can be installed by reversing the removal procedure.

CYLINDER HEAD

Model 40D4

45A. To remove the cylinder head, first drain cooling system and remove center hood. Note: If more working room is desired, first remove grilles, then remove front and center hood as a unit. Remove valve cover, and disconnect the rocker shaft oil feed pipe; then remove the rocker arms assembly and push rods. Disconnect the nozzle leak-off line, remove the high pressure lines connecting the injection pump to the injection nozzles and immediately cap the connections to prevent the entrance of dirt or other foreign material. Remove the injection nozzles. Unbolt the fuel line and linkage clips from left side of head. Disconnect the upper radiator hose and by-pass hose. Disconnect the heat indicator bulb and steering shaft support from rear of head. Disconnect air cleaner hose, vacuum line, throttle rod and heater cable from venturi and remove venturi and manifold. Remove the cylinder head retaining stud nuts and lift cylinder head from engine.

Cylinder head gasket is marked "FRONT" and "TOP" and should be coated with sealing compound before installation. Tighten the cylinder head retaining stud nuts in the sequence shown in Fig. CS32 and to a torque of 80-85 ft.-lbs. Recheck the nut torque after engine is hot. Intake and exhaust valve tappet gap is 0.010 hot.

Model 35

45B. To remove the cylinder head, first drain cooling system and remove center hood. Note: If more working room is desired, first remove grilles, then remove front and center hood as a unit. Remove the valve cover, rocker arms assembly and push rods. Unbolt carburetor from manifold and remove manifold. Disconnect the spark plug wires. Disconnect the heat indicator bulb and water outlet casting from head. Remove the cylinder head retaining stud nuts and lift cylinder head from engine.

Cylinder head gasket is marked "UP". Gasket should not be coated with sealing compound. Tighten the cylinder head retaining stud nuts in

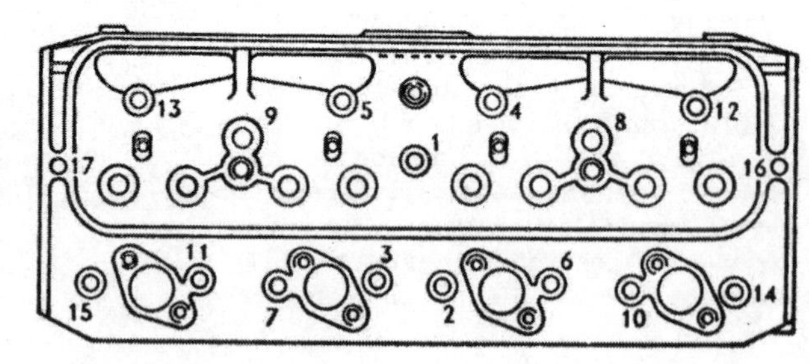

Fig. CS32 — Top view of model 40D4 cylinder head showing the nut tightening sequence. Specified torque value is 80-85 ft. lbs.

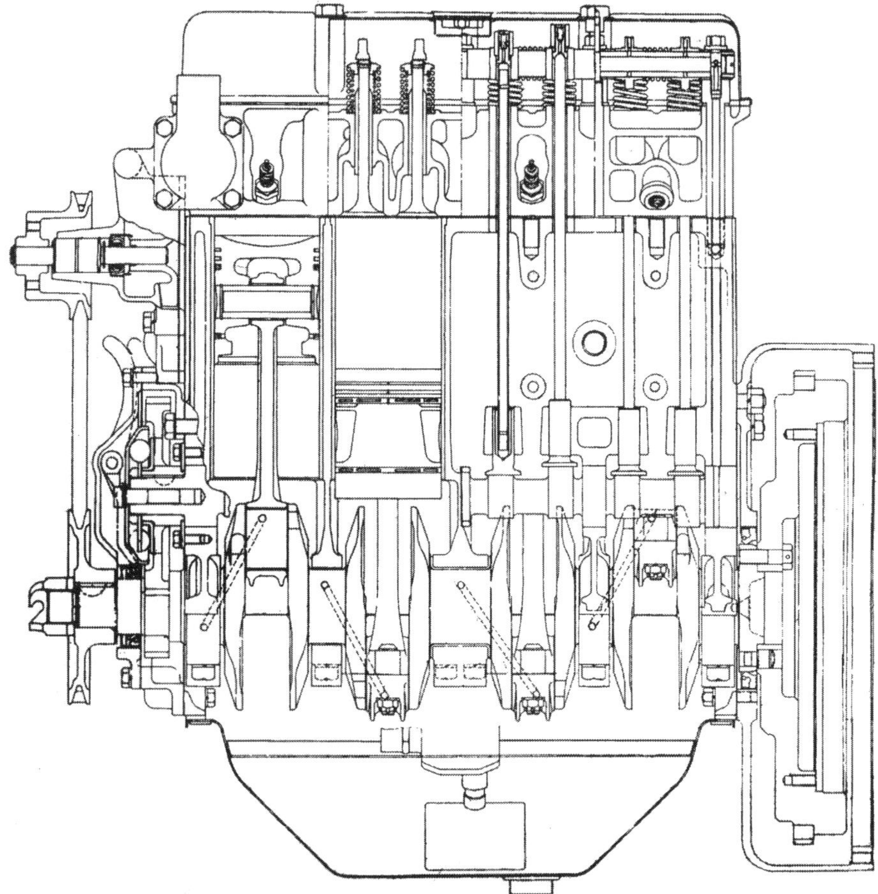

Fig. CS32A—Side sectional view of Hercules GO-198 engine typical of that used in Cockshutt model 35. The unit is equipped with a flyball type governor, precision insert type bearings and dry type sleeves. Electrical equipment is manufactured by Auto-Lite.

the sequence shown in Fig. CS33 and to a torque of 140 ft.-lbs. Recheck the nut torque after engine is hot. Intake and exhaust tappet gap is 0.010 hot.

VALVES AND SEATS

Model 40D4

47A. Intake and exhaust valves are not interchangeable and seat directly in cylinder head. Valve heads and cylinder head are numbered consecutively from front of engine. Refer to Fig. CS36. Any replacement valves should be so marked prior to installation. Intake and exhaust valves have a seat angle of 44 degrees, a face angle of 45 degrees and a desired seat width of $\frac{1}{16}$-inch. Seats can be narrowed, using 20 and 70 degree stones. Valves have a stem diameter of 0.3735-0.3745.

After valves are installed, check the clearance between head of valves and gasket surface of cylinder head, using a straight edge and a feeler gage as shown in Fig. CS36. Minimum allowable clearance is 0.057 for the intake, 0.053 for the exhaust. If clearance is less than specified, it will be necessary to reface the valve until at least the minimum clearance is obtained. Intake and exhaust valve tappet gap should be set to 0.010 hot.

Model 35

47B. Intake and exhaust valves are not interchangeable and seat directly in cylinder head. Valves have a face and seat angle of 45 degrees and a desired seat width of $\frac{5}{32}$-inch. Seats can be narrowed, using 20 and 70 degree stones. Valves have a stem diameter of 0.3725-0.3735.

Adjust the intake and exhaust tappet gap to 0.010 hot. Tappet adjusting screws are of the self-locking type.

VALVE GUIDES AND SPRINGS

Model 40D4

50A. New intake and exhaust valve guides are interchangeable. Guides can be pressed or driven from cylinder head if renewal is required. Press new guides into head until shoulder on guide seats against head. The 0.3735-0.3745 diameter valve stems should have a clearance of 0.0015-0.004 in the 0.3760-0.3775 diameter guides.

50B. Each valve is fitted with an inner and outer spring. Renew any spring which is rusted, discolored or does not meet the test specifications which follow:

Inner springs should test 30-31.5 lbs. when compressed to a height of $1\frac{5}{32}$ inches.

Outer springs should test 55-58.5 lbs. when compressed to a height of $1\frac{13}{32}$ inches.

Model 35

50C. New intake and exhaust valve guides are interchangeable. Guides can be pressed or driven from cylinder head if renewal is required and new guides should be installed with smaller O. D. of guide up. Distance from

Fig. CS33—Top view of model 35 cylinder head showing the nut tightening sequence. Specified torque value is 140 ft. lbs.

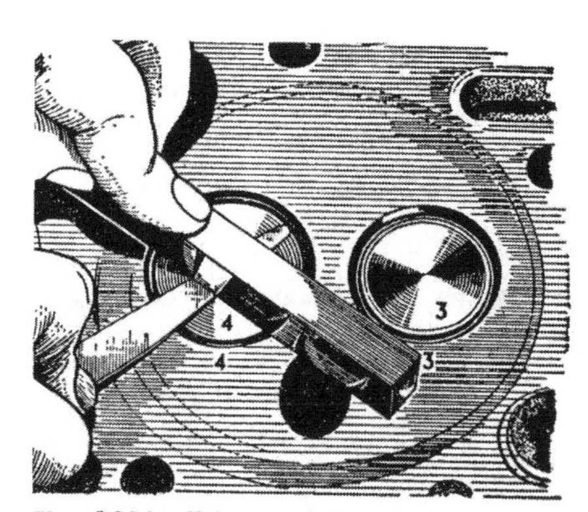

Fig. CS36—Using straight edge and feeler gage to check head clearance of valves on 40D4 engines. Notice also the number markings showing valve location.

port end of guides to gasket surface of cylinder head is 1⅜ inches. Ream new guides after installation to provide the recommended minimum stem to guide clearance of 0.0015-0.0025. Maximum allowable clearance is 0.006.

Refer to paragraph 47B for valve stem diameter.

50D. Exhaust valve springs are not interchangeable with either the inner or outer intake valve springs . Renew any spring which is rusted, discolored or does not meet the test specifications which follow:

Exhaust valve springs have a free length of approximately 1 57/64 inches and should test 80-86 lbs. when compressed to a height of 1 11/64 inches.

Intake valve inner springs have a free length of approximately 1⅞ inches and should test 32-36 lbs. when compressed to a height of $1\frac{3}{32}$ inches.

Intake valve outer springs have a free length of approximately $2\frac{7}{32}$ inches and should test 46-52 lbs. when compressed to a height of $1\frac{7}{32}$ inches.

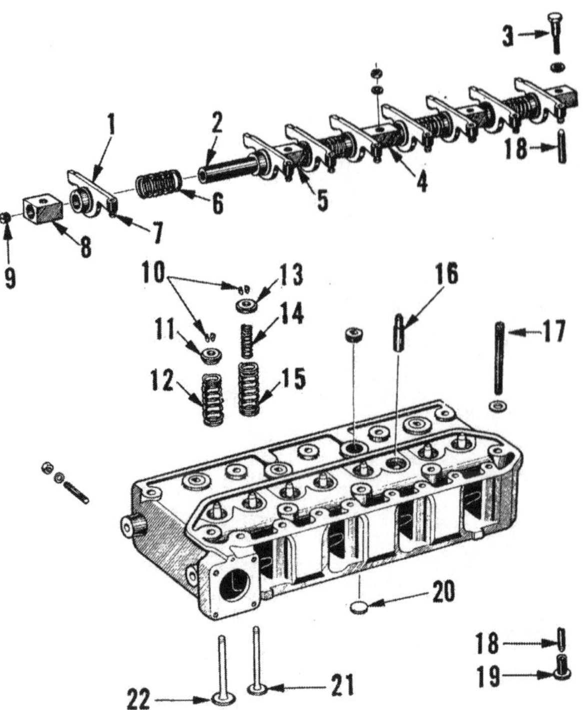

Fig. CS40B — Exploded view of model 35 cylinder head, rocker arms and valve mechanism. Exhaust valves are equipped with positive type rotators.

1. Rocker arm
2. Rocker shaft
3. Drilled screw
4. Shaft center support
5. Shaft intermediate support
6. Spring
7. Tappet adjusting screw
8. Shaft end support
9. Shaft end plug
10. Split cone keepers
11. Rotator
12. Exhaust valve spring
13. Spring retainer
14. Intake valve inner spring
15. Intake valve outer spring
16. Valve guide
17. Hollow stud
18. Push rod
19. Cam follower
20. Cup plug
21. Intake valve
22. Exhaust valve

VALVE TAPPETS *(Cam Followers)*

Model 40D4

53A. The 0.7475-0.7485 diameter mushroom type tappets have a clearance of 0.001-0.0032 in the 0.7495-0.7507 diameter cylinder block bores and are available in standard size only.

To remove the tappets, it is necessary to remove the camshaft as outlined in paragraph 64A. Adjust the valve tappet gap to 0.010 hot.

Model 35

53B. The 0.7485 - 0.7490 diameter mushroom type tappets operate directly in the unbushed cylinder block bores with a desired clearance of 0.0005-0.0015 and are available in standard size only. Maximum allowable clearance is 0.005.

To remove the tappets, it is first necessary to remove the camshaft as outlined in paragraph 64B.

When reassembling, adjust the tappet gap to 0.010 hot.

ROCKER ARMS

Model 40D4

54A. An exploded view of the cylinder head, rocker arms and associated parts is shown in Fig. CS40A. The procedure for removing and disassembling the rocker arms unit is evident. The 0.6223-0.6238 diameter rocker arm shaft has a clearance of 0.0007-0.0035 in the 0.6245-0.6258 diameter rocker arm bushings. Bushings are available separately for field installation or, re-

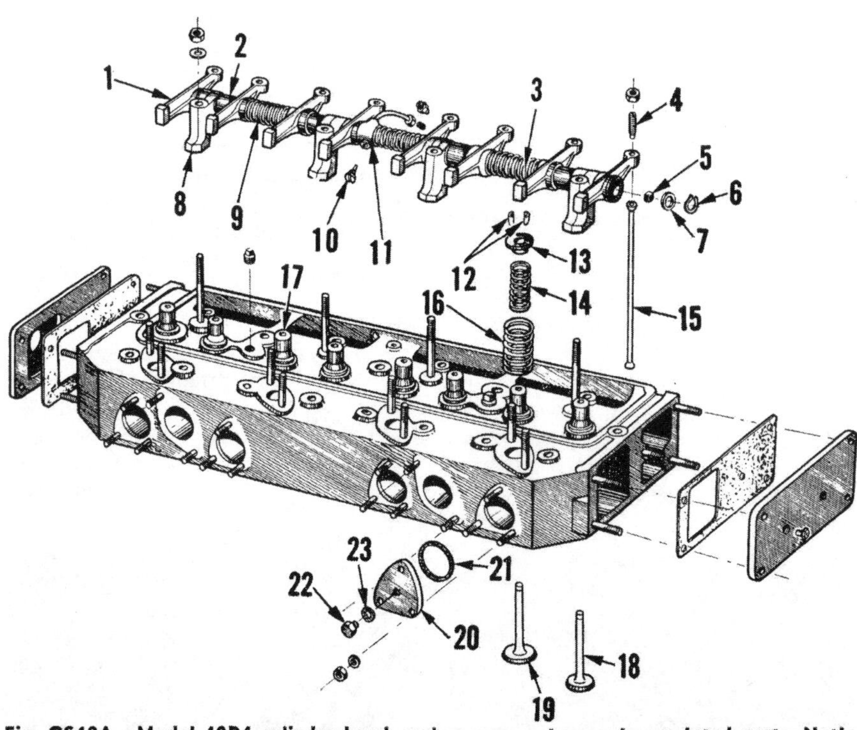

Fig. CS40A—Model 40D4 cylinder head, rocker arms, valves and associated parts. Notice that head is fitted with removable end covers. Rocker arm bushings are available separately for field installation.

1. Rocker arm and bushing
2. Spacer
3. Rocker shaft
4. Tappet adjusting screw
5. Shaft end plug
6. Retainer
7. Washer
8. Bracket
9. Spring
10. Oil pipe locating screw
11. Oil pipe
12. Split cone keepers
13. Spring retainer
14. Inner valve spring
15. Push rod
16. Outer valve spring
17. Valve guide
18. Exhaust valve
19. Intake valve
20. Precombustion chamber cap
21. Gasket
22. Chamber cap plug
23. Washer

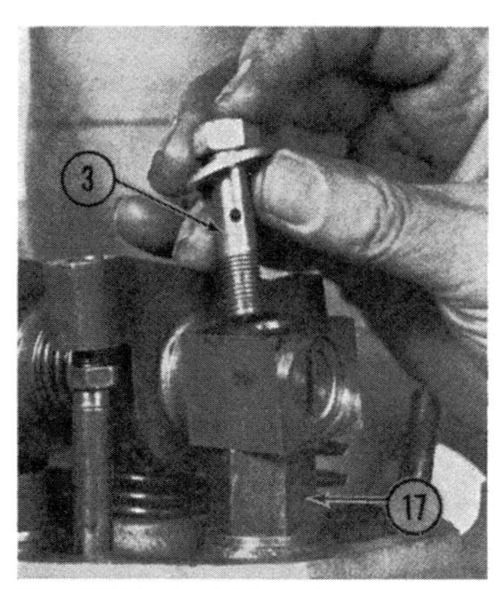

Fig. CS40C—Insufficient oil flow to model 35 rocker arms can be caused by foreign material in hollow stud (17) and/or screw (3).

placement rocker arms contain factory installed bushings. When installing bushings, make certain that oil hole in bushing is in register with oil hole in rocker arm and ream the bushings, if necessary, to obtain the desired clearance.

After engine is started, if oil does not flow from hole in top of each rocker arm, check for foreign material in the oil supply line.

Model 35

54B. An exploded view of the rocker arms and associated parts is shown in Fig. CS40B. All rocker arms are interchangeable and contain self-locking adjusting screws. Lubrication to the assembly is supplied through the hollow stud (17). If oil does not flow to the rocker arms, check for foreign material in hollow stud (17), screw (3) or the internal oil passages. Refer also to Fig. CS40C.

The 0.859-0.860 diameter rocker arm shaft has a normal clearance of 0.003 in the rocker arms. If clearance exceeds 0.006, renew the arms and/or shaft.

VALVE TIMING

Model 40D4

55A. Valves are properly timed when all timing marks are in register as shown in Fig. CS41A. Chisel marked line on camshaft must register with similar mark on camshaft gear hub. Chisel marked tooth on camshaft gear must mesh with chisel marked tooth space on idler gear and chisel marked tooth on crankshaft gear must mesh with chisel marked tooth space on idler gear.

Notice also that chisel marked tooth on injection pump drive gear meshes with chisel marked tooth space on idler gear.

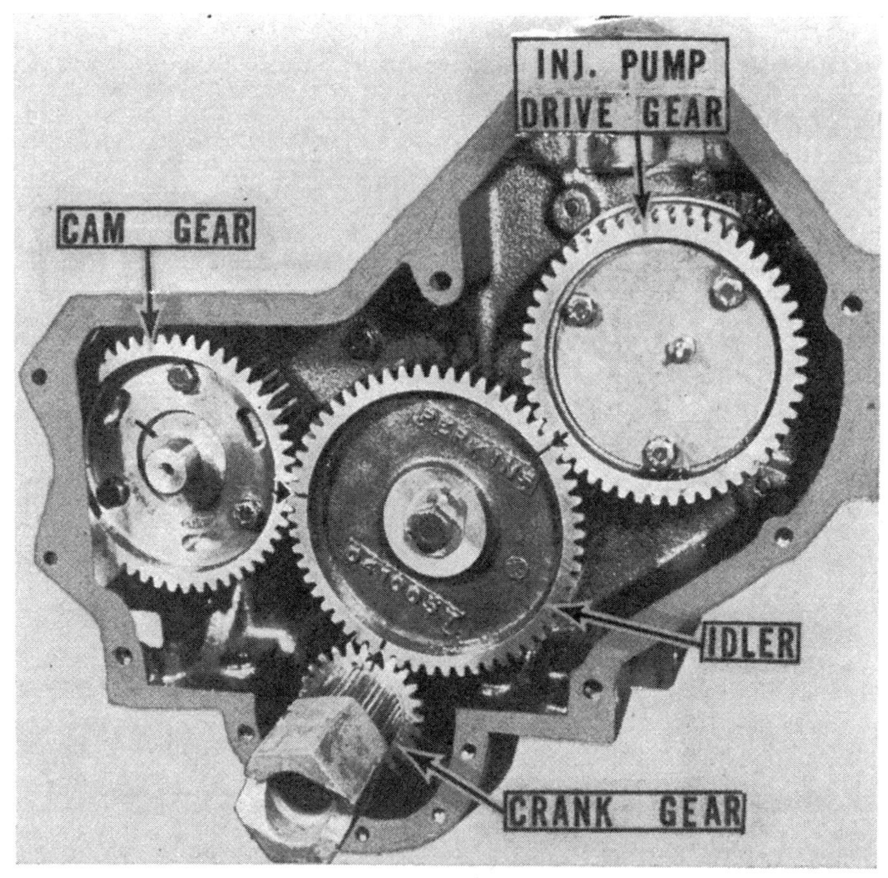

Fig. CS41A—Model 40D4 timing gear train. When installing the idler gear, make certain that all timing marks are in register as shown.

Model 35

55B. Valves are properly timed when the punch marked tooth on crankshaft gear is meshed with the punch marked tooth space on camshaft gear as shown in Fig. CS41B.

VALVE ROTATORS

Model 35

55C. Positive type exhaust valve rotators ("Rotocaps") are factory installed on model 35 tractors. Normal servicing of the rotators consists of renewing the units. It is important, however, to observe the valve action after engine is started. The valve rotator action can be considered satisfactory if the valve rotates a slight amount each time the valve opens.

Fig. CS41B — Model 35 timing gear train. Gears must be meshed so that the punched timing marks are in register.

CRANKCASE FRONT COVER

Model 40D4

57A. To remove the timing gear case cover, first drain cooling system and remove radiator as outlined in paragraph 195A; then, remove fan and fan belt. Remove batteries, battery boxes and clutch housing cover. Unbolt the timing gear case cover from engine frame and be careful not to mix or lose shims (41 and 42—Fig. CS49B) which may be installed between gear case cover and frame. Remove the engine rear mounting bolts and using a drift, bump both of the engine locating dowels up and out of the engine frame. Raise front of engine slightly and remove the crankshaft pulley. Remove the cap screws retaining cover to timing gear case and remove cover from engine. The crankshaft front oil seal (43) which is retained in the timing gear case cover can be renewed at this time. Install seal with lip of same facing inward.

Before installing timing gear case cover, remove the inspection cover (39) on which is mounted the engine hour meter. Install the timing gear case cover but do not tighten the retaining cap screws until after the crankshaft pulley is installed and the crankshaft turned several revolutions. This procedure facilitates locating the front oil seal with respect to the pulley hub. Tighten the cover retaining cap screws and install the inspection cover (39) making certain that lug on hour meter engages pin on drive plate fastened to the injection pump drive gear.

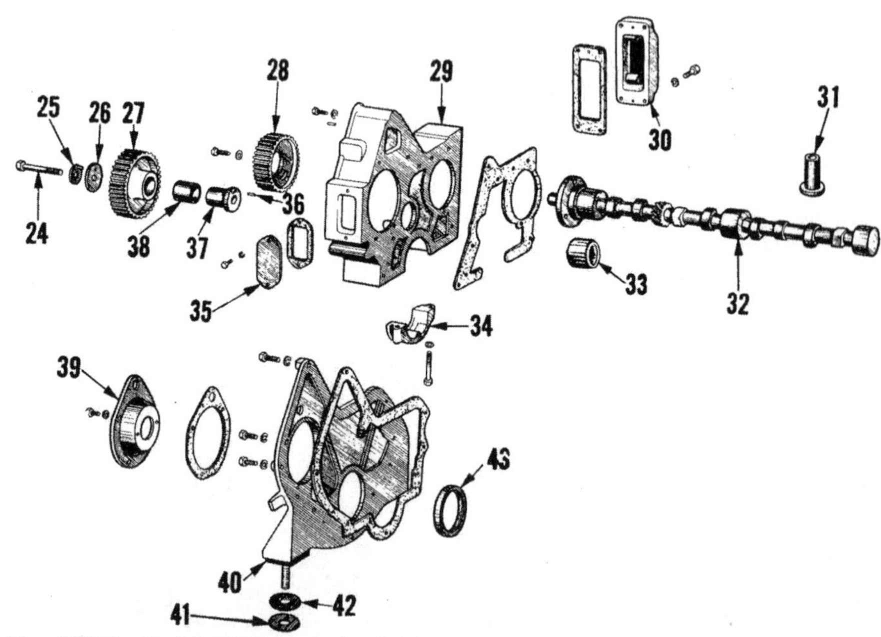

Fig. CS49B—Model 40D4 camshaft, timing gear case, cover and associated parts. Camshaft front bushing (33) is used on models after Engine Ser. No. 6018995.

24. Idler gear retaining screw
25. Washer
26. Plate
27. Idler gear
28. Cam gear
29. Timing gear case
30. Right hand cover
31. Cam follower
32. Camshaft
33. Bushing
34. Bottom cover
35. Inspection cover
36. Dowel
37. Idler gear shaft
38. Bushing
39. Front inspection cover
40. Timing gear case cover
41. Shims (0.005)
42. Shims (0.010)
43. Oil seal

Model 35

57B. To remove the timing gear housing cover, first drain cooling system and remove radiator as outlined in paragraph 195A; then remove fan and fan belt.

Loosen the water pump retaining screws and, using a suitable puller, remove the crankshaft pulley. Disconnect controls from governor lever, then unbolt and remove the timing gear case cover. Note: Removal of cover will be made easier if governor lever is held forward as cover is withdrawn.

The crankshaft front oil seal (47—Fig. CS49C) which is retained in the timing gear case cover can be renewed at this time. Install seal with lip of same facing inward.

Before installing the timing gear housing cover, make certain that governor flyballs and outer race (46) are properly positioned as shown in Fig.

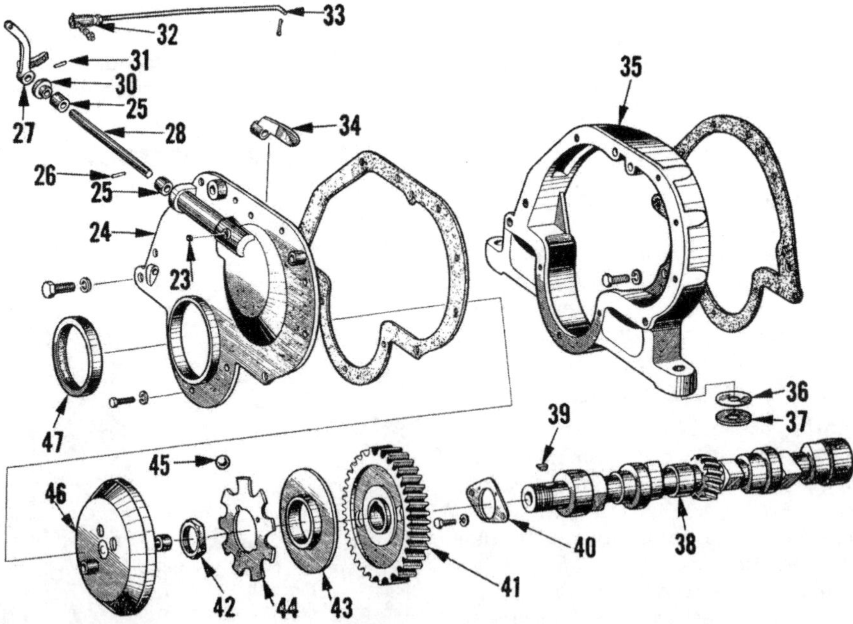

Fig. CS49C—Model 35 camshaft, timing gear case, cover, governor components, etc. Notice that governor weight unit is mounted on cam gear.

23. Cup plug
24. Timing gear case cover
25. Governor shaft bushings
26. Governor lever pin
27. Outer governor lever
28. Governor shaft
30. Seal
31. Pin
32. Ball joint
33. Governor to carburetor rod
34. Inner governor lever
35. Timing gear case
36. Shim (0.010)
37. Shim (0.005)
38. Camshaft
39. Woodruff key
40. Thrust plate
41. Cam gear
42. Nut
43. Inner race
44. Ball driver
45. Flyball
46. Outer race
47. Oil seal

CS49D. Heavy grease will facilitate holding the balls in position as shown. Install the timing gear housing cover but do not tighten the retaining screws until after the crankshaft pulley is installed and the crankshaft turned several revolutions. This procedure facilitates locating the front oil seal with respect to the pulley hub.

TIMING GEARS

Model 40D4

60A. Timing gear train consists of the crankshaft gear, camshaft gear, idler gear and injection pump drive gear as shown in Fig. CS41A. Normal timing gear backlash is 0.003-0.006. If backlash is excessive, renew the worn gear or gears. To remove the camshaft gear, crankshaft gear and/or idler gear, first remove the timing gear case cover as outlined in paragraph 57A and proceed as in the following paragraphs. The injection pump drive gear can also be removed when timing gear case cover is off, but if the injection pump drive gear is the only one to be renewed, the quickest method is to remove the injection pump as outlined in paragraph 156A.

60B. **IDLER GEAR.** To remove the idler gear, remove cap screw (24—Fig. CS49B), washer (25) and plate (26). Withdraw the idler gear and bushing unit from the idler gear shaft. The 1.497-1.498 diameter idler gear shaft has an operating clearance of 0.001-0.003 in the 1.499-1.500 diameter bushing. If clearance is excessive, renew the pre-sized bushing and/or shaft. When installing idler gear, mesh the gears so that timing marks are in register as shown in Fig. CS41A. Examine plate (26—Fig. CS49B) and renew same if it is worn on rear face.

60C. **CAMSHAFT GEAR.** To remove camshaft gear, first remove idler gear and proceed as follows: Remove the small cover (30—Fig. CS49B) from right side of timing gear case. Remove the three cap screws retaining gear to camshaft and remove the gear. When installing the camshaft gear, make certain that the chisel mark on gear hub is in register with chisel mark on camshaft and tighten the retaining screws securely. When installing the idler gear, mesh the gears so that timing marks are in register as shown in Fig. CS41A.

60D. **CRANKSHAFT GEAR.** To remove the crankshaft gear, first remove the idler gear and proceed as follows: Drain the lubricating oil and remove oil pan. Remove the timing gear case bottom cover (34—Fig. CS49B) and using a suitable puller, remove the crankshaft gear. To facilitate reinstallation of the gear, boil same in oil for about 15 minutes and using a brass drift, bump gear on crankshaft. When installing the idler gear, mesh the gears so that timing marks are in register as shown in Fig. CS41A.

60E. **PUMP DRIVE GEAR.** To remove the injection pump drive gear when timing gear cover is off, crank engine until number one piston is coming up on compression stroke and continue cranking until the "DC" line on flywheel is in register with pointer which can be viewed through inspection port in left side of clutch housing cover. Note: The "DC" mark is stamped on rear face of flywheel and can be viewed, using an inspection mirror.

At this time, the "S" marked line on the fuel injection pump gear adapter will be in register with the scribed line on pointer extending from front face of pump as shown in Fig. CS50B. Now, unbolt and remove idler gear and injection pump drive gear.

Reinstall idler gear so that timing marks on the meshing gear teeth of camshaft gear, crankshaft gear and idler gear are in register as shown in Fig. CS41A. Install the injection pump drive gear so that timing marks on idler gear and pump drive gear are in register and tighten the retaining screws finger tight.

Grasp the protruding rear end of the injection pump camshaft with a pair of pliers, turn the shaft slightly to again align the "S" marked line on the fuel injection pump gear adapter with the scribed line on the pointer extending from front face of pump as shown in Fig. CS50B and while holding the pump camshaft in this position, tighten the three cap screws retaining the pump drive gear to the injection pump gear adapter.

Model 35

60F. Timing gear train consists of the camshaft gear and crankshaft gear as shown in Fig. CS41B. Normal timing gear backlash is 0.000-0.002. If backlash exceeds 0.008, it will be necessary to renew the worn gear (or gears). To remove the gears, proceed as follows:

60G. **CAMSHAFT GEAR.** To remove the camshaft gear, first remove the timing gear housing cover as outlined in paragraph 57B. Remove nut (42—Fig. CS49C) from end of camshaft and withdraw the governor ball driver (44) and inner race (43). Then, using a suitable puller, remove camshaft gear from shaft.

When reassembling, mesh the gears so that timing marks are in register as shown in Fig. CS41B. Install the governor inner race and ball driver on camshaft gear hub, then install and

Fig. CS49D — On model 35, use heavy grease to hold the governor flyballs in position when installing the outer race.

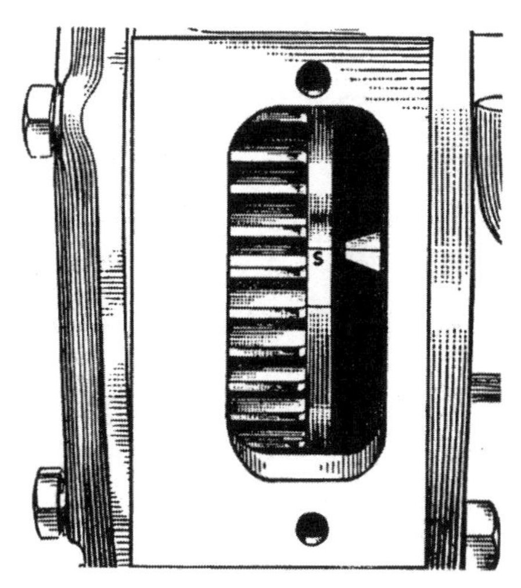

Fig. CS50B—Model 40D4. "S" marked line on injection pump drive gear and scribed line on pointer in register.

tighten the gear retaining nut to a torque of 130 ft.-lbs. Check to be sure the inner race rotates freely on the gear hub and that the race has a fore and aft end play of 0.007-0.020. If end play is excessive, renew the race and ball driver and recheck. If end play is insufficient, file off the necessary amount of metal from the race hub.

60H. **CRANKSHAFT GEAR.** The crankshaft gear is heated prior to installation and therefore has an extremely tight fit on the crankshaft. To renew the gear without removing crankshaft, first remove the camshaft gear as outlined in paragraph 60G and proceed as follows:

Remove oil pan and remove the cap screws retaining the oil pan front adapter to timing gear housing. Unbolt timing gear housing from engine frame and be careful not to mix or lose shims (36 and 37—Fig. CS49C) which may be installed between gear housing and frame. Remove the clutch housing top cover. Remove the engine rear mounting bolts and using a drift bump both of the engine locating dowels up and out of the engine frame. Raise front of engine slightly, then unbolt and remove the timing gear housing. At this time the crankshaft gear can be removed by using a heavy puller or the gear can be split and removed as follows:

Using a ¼-inch diameter drill centered between edge of keyway and base of gear teeth, drill through the gear and spread gear with a chisel. Be careful, however, not to damage any other parts during the drilling operation.

To facilitate reinstallation of the gear, boil same in oil for about 15 minutes or heat gear evenly with a torch until gear turns a pale straw yellow color and using a brass drift, bump gear on crankshaft. When reassembling, install camshaft gear as outlined in paragraph 60G.

CAMSHAFT AND BEARINGS

Model 40D4

64A. On engines prior to Serial No. 6018996, the three camshaft bearing journals ride directly in the unbushed cylinder block bores. On engines after Serial No. 6018995, the two rear camshaft bearing journals ride directly in the unbushed cylinder block bores; whereas, the front journal is carried in a renewable, pre-sized bushing. To remove the camshaft, first remove the timing gear cover as outlined in paragraph 57A; then remove the valve cover, rocker arms assembly and push rods. Remove oil pan and oil pump. Remove cover (30—Fig. CS49B) from right side of timing gear case and remove idler gear (27). Push the tappets up into their bores, withdraw camshaft and gear unit from cylinder block, then unbolt and remove gear from shaft.

Fig. CS51A — On model 35, the camshaft retaining thrust plate is retained to front face of cylinder block by cap screws (CS).

Check the camshaft bearing journals against the values which follow:

	Journal Diameter
No. 1 (front)	2.0565-2.0575
No. 2	2.0465-2.0475
No. 3	2.0365-2.0375

Check the camshaft bearing bores against values which follow:

	Bore Diameter
No. 1 (front)	2.0615-2.0635
No. 2	2.0515-2.0535
No. 3	2.0415-2.0435

Normal clearance of camshaft bearing journals in bores is 0.004-0.007. If clearance is excessive, renew the worn part.

When installing the camshaft, reverse the removal procedure and make certain the valve timing marks are in register as shown in Fig. CS41A. Normal camshaft end play of 0.021-0.038 is non-adjustable.

Model 35

64B. The four camshaft bearing journals ride directly in the unbushed cylinder block bores with a normal clearance of 0.0015-0.0035. To remove the camshaft, first remove the timing gear cover as outlined in paragraph 57B; then remove the valve cover, rocker arms assembly and push rods. Remove the ignition distributor, oil pan and oil pump. Withdraw the governor outer ball race, then remove nut and withdraw the ball driver and inner race as shown in Fig. CS51A. Working through openings in camshaft gear, remove the camshaft thrust plate retaining cap screws (CS), push tappets up into their bores and withdraw camshaft and gear unit from cylinder block. Note: Home made wires similar to that shown in Fig. CS51B can be inserted down through push rod openings in cylinder head to hold the tappets in the raised position.

64C. Before removing gear from camshaft, insert a feeler gage between the thrust plate and gear and check the camshaft end play which should be 0.002-0.007. If end play exceeds 0.025, and if the thrust plate is worn, renew the thrust plate; if a new thrust plate does not reduce the end play to within the desired limits, it will be necessary to remove, by draw filing, the necessary amount of metal from the rear end of the gear hub on which the thrust plate rides. If the end play is insufficient, it will be necessary to either reduce the thickness of the thrust plate or make up and add a shim between the gear hub and the camshaft.

The 2.053-2.054 diameter camshaft bearing journals have a normal operating clearance of 0.0015-0.0035 in the 2.0555-2.0565 cylinder block bores. If clearance exceeds 0.006, it will be necessary to renew camshaft and/or cylinder block.

When installing the camshaft, reverse the removal procedure and mesh the timing gears as shown in Fig. CS-41B. Tighten the camshaft gear nut to torque of 130 ft.-lbs. and refer to paragraph 193 when installing the governor components. Install oil pump as outlined in paragraph 78E and the

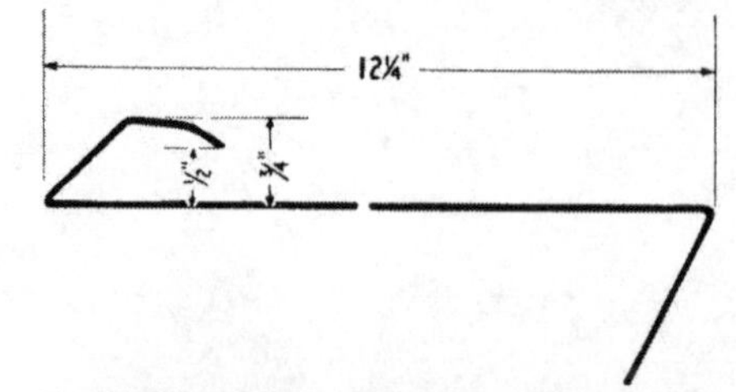

Fig. CS51B—On model 35, wires can be bent as shown to hold the valve tappets (cam followers) in the raised position to facilitate withdrawing the camshaft.

ignition distributor as outlined in paragraph 201A. Adjust the tappets to 0.010 hot.

CONNECTING RODS AND PISTON UNITS

Model 40D4

65A. Piston and connecting rod units are removed from above after removing cylinder head as in paragraph 45A and oil pan as in paragraph 76A.

Cylinder numbers are stamped on connecting rod and cap. When reinstalling the rod and piston units, make certain that numbers are in register and face toward camshaft side of engine. Tighten the connecting rod bolts to a torque of 100-105 ft. lbs.

Model 35

65B. Piston and connecting rod units are removed from above after removing the cylinder head as outlined in paragraph 45B and the oil pan as outlined in paragraph 76B.

Cylinder numbers are stamped on connecting rod and cap. When reinstalling the rod and piston units, make certain that numbers are in register and face toward camshaft side of engine. Tighten the connecting rod bolts to a torque of 56 ft.-lbs.

PISTONS, RINGS AND SLEEVES

Model 40D4

68A. As shown in Fig. CS52A, each piston is fitted with one chrome plated compression ring, one plain compression ring, one laminated compression ring (four laminations) and two oil control or scraper rings. On engines used in early tractor, the top ring is of the chrome plated type and the second ring is plain; whereas on later engines, the top groove is fitted with a plain compression ring and the second ring is of the chrome plated type.

Check the pistons and rings against the values which follow:

Top comp., end gap......0.017 -0.022
Second comp., end gap...0.012 -0.017
Oil (Scraper), end gap...0.012 -0.017
Comp., side clearance....0.002 -0.004
Oil (Scraper), side clearance........0.0025-0.0045

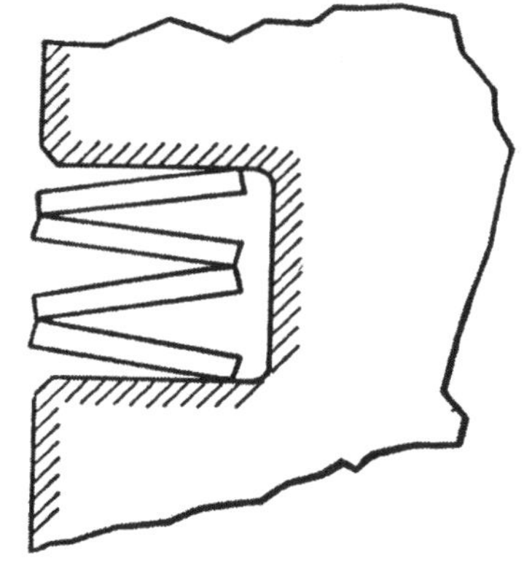

Fig. CS52B—On model 40D4, the four sections of the laminated compression ring should be installed in the third groove as shown.

Install the laminated compression ring as shown in Fig. CS52B.

Cylinders are fitted with renewable wet type sleeves which have a bore diameter of 4.250-4.251. Sleeves should be renewed if out-of-round exceeds 0.010 and/or if taper exceeds 0.006. Desired clearance between piston skirt and cylinder sleeve is 0.004-0.006.

When connecting rod and piston units are installed, check the amount each piston stands out above the cylinder block as follows: With the pistons in TDC position, make certain the sleeves are completely down, then using a straight edge and feeler gage as shown in Fig. CS52C, check and record the amount of sleeve stand-out above the cylinder block. Then check and record the amount the sleeve stands-out above top of piston as shown in Fig. CS52D. Now, by subtracting the second value (sleeve above piston) from the first value (sleeve above block), you have determined the amount the piston stands-out above the cylinder block. If stand-out is less than 0.007, it is recommended that piston be discarded and a new one installed and checked. If stand-out is more than 0.012, remove the piston, mount same in a lathe and machine off the necessary amount from the piston head. Refer to paragraph 69A for R&R of sleeves.

LATE MODEL
Plain Compression
Chrome Compression
Laminated Compression
Slotted Scraper Ring

EARLY MODEL
Chrome Compression
Plain Com`ression
Laminated Compression
Slotted Scraper Ring

Fig. CS52A—Model 40D4 piston, showing ring location for both early and late construction. On early models, the chrome plated compression ring is in the top groove. On late models, the chrome plated ring is in the second groove.

Model 35

68B. Pistons are fitted with two compression rings and two oil control rings which have an end gap of 0.010-0.020 and a side clearance of 0.0035-0.005 for the top compression ring, 0.002-0.0035 for the second compression ring, 0.0015-0.003 for the oil control rings.

Cylinders are fitted with renewable dry type sleeves which have a bore diameter of 3.7495-3.7505. Sleeves should be renewed if out-of-round exceeds 0.010 and/or if taper exceeds 0.005. Desired clearance between piston skirt and cylinder sleeve is 0.003-0.0035. Maximum allowable clearance is 0.012. Piston to sleeve clearance can be considered satisfactory if a spring scale pull of 5-7 lbs. will withdraw a 0.003 thick, ½-inch wide feeler ribbon.

Refer to paragraph 69B for R&R of sleeves.

R&R CYLINDER SLEEVES

Model 40D4

69A. The wet type cylinder sleeves can be renewed after removing the connecting rod and piston units. When sleeves are renewed, it is recommended that pistons also be renewed. Coolant leakage at bottom of sleeves is prevented by two rubber

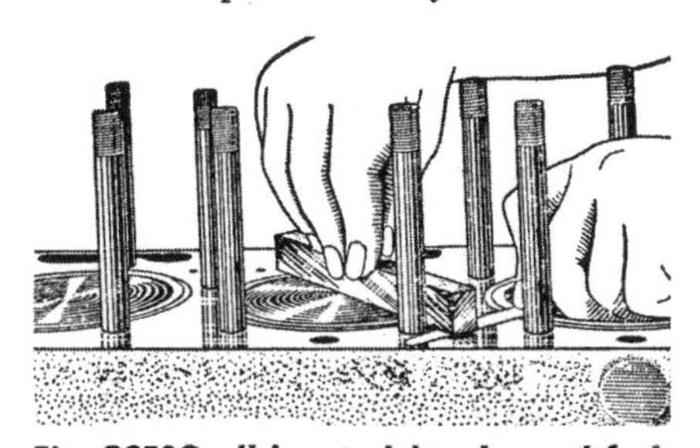

Fig. CS52C—Using straight edge and feeler gage to check the amount the 40D4 cylinder sleeves stand out above the cylinder block.

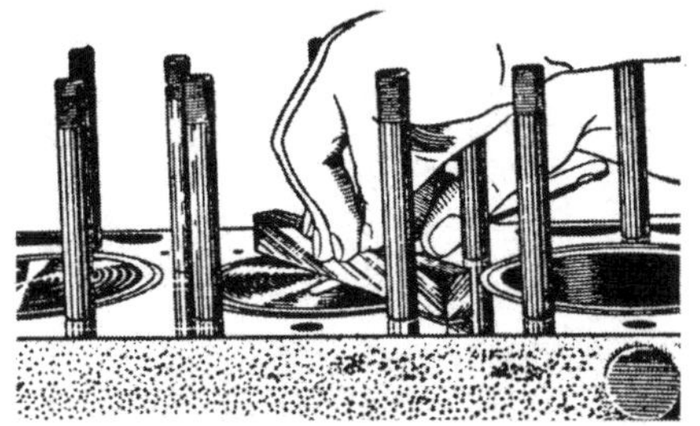

Fig. CS52D—Using straight edge and feeler gage to check the amount the 40D4 cylinder sleeves stand out above the pistons.

sealing rings. The cylinder block is counter-bored at the top to receive the sleeve flange and the head gasket forms the coolant seal at this point. The sleeves can be removed, using a special puller; or, the sleeves can be removed by placing a wood block against bottom edge of sleeve and tapping the block with a hammer.

Before installing new sleeves, thoroughly clean the cylinder block, paying particular attention to the seal seating surfaces at bottom and the counterbore at top. Using care not to stretch or twist the sealing rings, install the rings, coat same with brake fluid or a thick soap solution and press the sleeves in position with finger pressure only. If undue pressure is required to install sleeves, remove same and check for possible interferring foreign material.

Model 35

69B. The dry type cylinder sleeves can be renewed after removing the connecting rod and piston units. When sleeves are renewed, it is recommended that pistons also be renewed. Sleeves are a very light press fit in cylinder block and removal of same does not normally necessitate the use of a puller. If difficulty is encountered, place a wood block against bottom of sleeve and tap the block with a hammer.

Before installing new sleeves, be sure to thoroughly clean the cylinder block bores. Coat exterior of new sleeves with light engine oil and push the sleeves in place. It sometimes may be necessary to tap the sleeves downward, using a wood block and hammer.

Sleeves do not require any final sizing after installation.

PISTON PINS

Model 40D4

70A. The 1.4375-1.4385 diameter full floating type piston pins are retained in the piston bosses by snap rings and are available in standard size only. To facilitate removal and reinstallation of pin, heat piston in oil to approximately 110 degrees F. When installing piston pin bushing in connecting rod, be sure to align oil hole in bushing with oil hole in rod and ream the bushing to provide a diametral clearance of 0.0004-0.0018 for the pin.

Assemble No. 1 piston to No. 1 connecting rod, etc.

Model 35

70B. The 1.1245-1.1247 diameter full floating type piston pins are retained in the piston bosses by snap rings and are available in standard size as well as oversizes of 0.003 and 0.005. When installing piston pin bushing in connecting rod, be sure to align oil hole in bushing with oil hole in rod and ream the bushing to provide a diametral clearance of 0.0005 - 0.0012 for the pin. Pin should have a hand push fit or 0.0000-0.0005 clearance in piston.

CONNECTING RODS AND BEARINGS

Model 40D4

71A. Connecting rod bearings are of the non-adjustable, slip-in precision type renewable from below after removing the oil pan and rod bearing caps.

When installing new bearing shells, make certain that the bearing shell projections engage the milled slot in connecting rod and cap and that cylinder numbers on the rod and cap are in register and face toward camshaft side of engine. Bearing inserts are available in standard size as well as undersizes of 0.010, 0.020 and 0.030.

Check the crankshaft crankpins and the bearing inserts agaist the values which follow:

Crankpin diameter (Std.) . . 2.748-2.749

Bearing running
clearance 0.0025-0.0048

Rod side play 0.008-0.017

Rod bolt torque . . . . 100-105 ft.-lbs.

Model 35

71B. Connecting rod bearings are of the non-adjustable, slip-in precision type renewable from below after removing the oil pan and rod bearing caps.

When installing new bearing shells, make certain that the bearing shell projections engage the milled slot in connecting rod and cap and that cylinder numbers on the rod and cap are in register and face toward camshaft side of engine. Bearing inserts are available in standard size as well as undersizes of 0.010, 0.020, 0.030 and 0.040.

Check the crankshaft crankpins and the bearing inserts against the values which follow:

Crankpin diameter (Std.) . . 1.987-1.988

Bearing running
clearance (desired) 0.001-0.003

Bearing Running
clearance (max.) 0.006

Rod side play (desired) 0.005-0.012

Rod side play (max.) 0.020

Rod bolt torque . . . . 56 ft.-lbs.

CRANKSHAFT AND MAIN BEARINGS

Model 40D4

73A. Crankshaft is supported in three non-adjustable, slip-in precision type main bearings. Number one and two main bearing inserts can be removed from below in a conventional manner after removing oil pan, oil pump and main bearing caps. To remove the number three (rear) main bearing after oil pan is off, proceed as follows:

Remove the engine clutch as outlined in paragraph 218A. Remove the timing pointer, then unbolt and remove flywheel from crankshaft. Loosen the cap screws retaining the crankshaft rear oil seal housing to the cylinder block and remove the screws retaining the seal housing to the rear main bearing cap. The rear main bearing cap, bearing inserts and thrust washers can now be removed from below in the conventional manner.

Bearing inserts are available in standard size as well as undersizes of 0.010, 0.020 and 0.030. Main bearing caps and crankcase are numbers 1, 2, 3 and are located by offset dowels to prevent improper installation.

Normal crankshaft end play of 0.0045-0.0115 is controlled by thrust washers (52—Fig. CS55A) located on each side of the rear main bearing. If end play is not as specified, renew the thrust washers and recheck. Washers are available in standard size as well as 0.007 oversize.

To remove the crankshaft, remove engine, oil pan, oil pump, rod bearing caps, timing gear cover, idler gear, timing case lower cover, clutch, flywheel, engine rear support plate, rear oil seal retainer and main bearing caps.

Check the crankshaft and main bearings against the values which follow:

Crankpin diameter (Std.) . . 2.748-2.749

Main journal
diameter (Std.) 2.998-2.999

Main bearing
running clearance 0.0025-0.0048

Crankshaft end play 0.0045-0.0115

Main bearing
bolt torque . . . . 125-130 ft.-lbs.

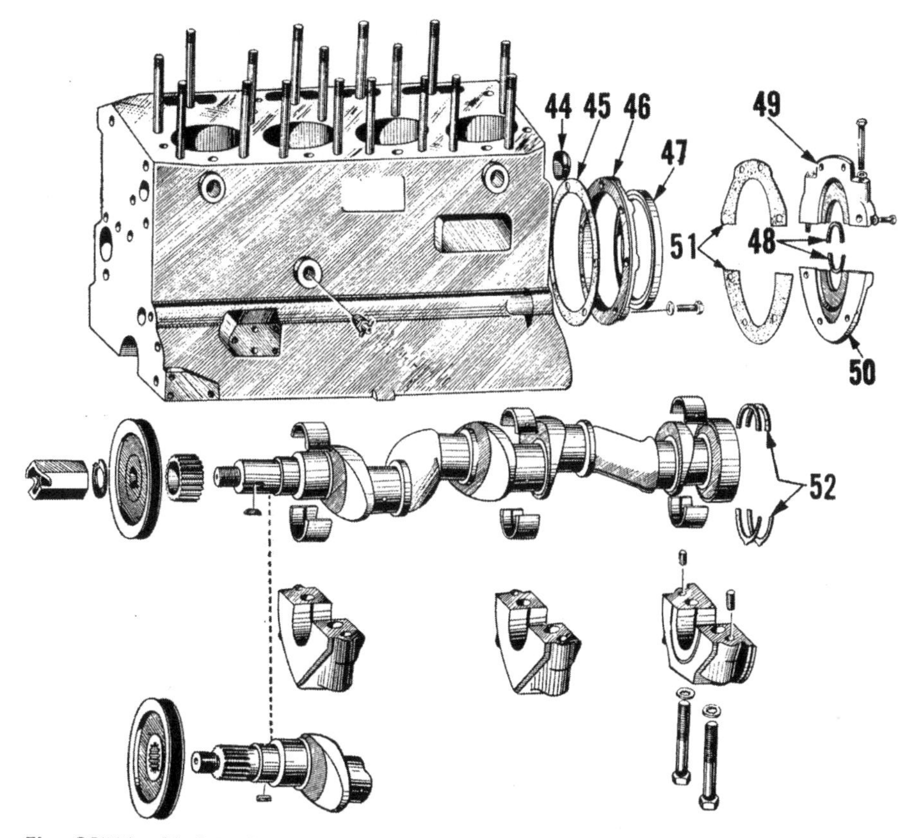

Fig. CS55A—Model 40D4 crankshaft, main bearings and rear oil seal. The one piece rear oil seal (47) is used on engines prior to Ser. No. 6030431 only.

44. Cup plug
45. Gasket, early models
46. Seal housing, early models
47. Seal, early models
48. Seal, late models
49. & 50. Seal housing, late models
51. Gasket, late models
52. Thrust washers

Model 35

73B. Crankshaft is supported in five non-adjustable, slip-in precision type main bearings. Bearing inserts can be renewed after removing oil pan, oil pump, front and rear oil pan adapters and main bearing caps. Bearing inserts are available in standard size as well as undersizes of 0.010, 0.020, 0.030 and 0.040. Normal crankshaft end play of 0.005-0.010 is controlled by the flanged center main bearing inserts. Maximum allowable end play is 0.020.

To remove the crankshaft, remove engine, oil pan, oil pump, rod bearing caps, timing gear housing cover, camshaft gear, timing gear housing, clutch, flywheel, engine rear end plate and main bearing caps.

Check the crankshaft and main bearings against the values which follow:

Crankpin diameter (Std.)..1.987-1.988
Main journal
diameter (Std.) 2.4965-2.4975
Main bearing running
clearance (desired) 0.0009-0.003
Main bearing running
clearance (Max.) 0.006
Crankshaft end play
(desired) 0.005-0.010
Crankshaft end play (Max.).....0.020
Main bearing bolt torque
Center ....70 ft.-lbs.
Front, rear &
intermediate ....80 ft.-lbs.

CRANKSHAFT REAR OIL SEAL

Model 40D4

75A. On engines prior to Serial No. 6030431, the crankshaft rear oil seal (47—Fig. CS55A) is a one-piece type located in housing (46) which is bolted to rear face of cylinder block. The procedure for removing the seal is evident after removing the flywheel as outlined in paragraph 77A.

On engines after Serial No. 6030430, the crankshaft rear oil seal (48—Fig. CS55A) is a two-piece type located in housings (49 and 50) which are bolted to rear face of cylinder block. To remove the seal, first remove the engine as outlined in paragraph 42A. Then remove the clutch, timing pointer and flywheel. Disconnect the necessary linkage and fuel lines and remove the engine rear support plate. The remaining procedure is evident.

Model 35

75B. Crankshaft rear oil seal is a one-piece spring loaded type, retained in the engine rear support plate. The procedure for removing the seal is evident after removing the flywheel as outlined in paragraph 77B.

OIL PAN

Model 40D4

76A. To remove the oil pan, drain the oil and remove the dip stick and dip stick tube. Loosen the oil filter mounting screws and remove the strainer cover and strainer from bottom of pan. The remainder of the removal procedure is conventional.

When installing the oil pan, be sure the front and rear arch gaskets are properly positioned before tightening the retaining screws.

Model 35

76B. The procedure for removing the oil pan is conventional. When installing the oil pan, use extra care to make certain that the front and rear arch gaskets do not slip out of position.

FLYWHEEL AND RING GEAR

Model 40D4

77A. To remove the flywheel, first remove the clutch as outlined in paragraph 218A. Remove the timing pointer then unbolt and remove flywheel from crankshaft.

The flywheel ring gear which is retained to front face of flywheel by cap screws can be renewed at this time. When reassembling, tighten the flywheel retaining cap screws to a torque of 75 ft.-lbs.

Model 35

77B. To remove the flywheel, first remove the clutch as outlined in paragraph 218A. Remove the safety wire, then unbolt and remove flywheel from crankshaft.

The flywheel ring gear can be removed by drilling and splitting same with a cold chisel. To facilitate installation of the gear, boil same in oil for fifteen minutes or heat the ring gear evenly with a torch.

OIL PUMP

Model 40D4

78C. The engine oil pump can be removed after removing the oil pan, disconnecting the delivery pipe from crankcase, removing the set screw retaining the suction pipe to the center main bearing cap and removing the pump retaining screw located on right side of crankcase. To disassemble the pump, remove the suction and delivery pipes, bottom cover (63—Fig. CS69), housing (60) and idler gear. Press off the spiral drive gear (53), withdraw the drive shaft and press off the driven gear. It is not necessary to remove the idler gear shaft unless replacement is required. Check the pump parts against the values which follow:

Gear bore diameter
in housing 1.339-1.341
Gear diameter 1.331-1.333
Diametral clearance of gears
in housing 0.006-0.010
Gear length 0.995-0.997
Depth of housing
gear pocket 0.998-1.001
Gear end clearance. 0.001-0.006
Drive shaft diameter. 0.501-0.5015
Drive shaft bushing dia. 0.502-0.503
Diametral clearance of shaft
in bushings 0.0005-0.002
Idler gear shaft dia. 0.4985-0.499
Idler gear bore 0.4998-0.5003
Diametral clearance of shaft in
idler gear 0.0008-0.0018

When reassembling, press driven gear on the shaft until end of gear is flush with end of shaft. Install drive shaft, then press the spiral drive gear on the shaft until end of gear is flush with end of shaft. Install the remaining pump components. NOTE: Pump relief valve must be on delivery side of pump. Be sure to lock the banjo bolts with safety wire.

Model 35

78D. To remove the engine oil pump, first remove the ignition distributor and oil pan, then unbolt and withdraw the oil pump from cylinder block. To disassemble the pump, remove pin (48—Fig. CS70) and using a suitable puller, remove the spiral drive gear from the pump shaft. Unbolt and remove pump cover (57) and withdraw the pumping gears and shafts. Gears can be pressed from the drive shaft and idler shaft if renewal is required. Check the pump components against the values which follow:

Gear bore diameter
in housing 1.5005-1.5015
Gear diameter 1.4975-1.4985
Diametral clearance of gears
in housing 0.002-0.003
Gear length 1.061-1.062
Depth of housing
gear pocket 1.064-1.065
Gear end clearance. 0.002-0.004
Drive shaft and idler
shaft diameter 0.6240-0.6245
Drive shaft and idler shaft bore
diameter in housing . .0.6255-0.6265

When reassembling, reverse the disassembly procedure and press the spiral drive gear on, to the dimension shown in Fig. CS71.

78E. When installing the oil pump, crank engine until number one piston is coming up on compression stroke and continue cranking until the "DC" line on flywheel rim is aligned with notch in flywheel housing as shown in Fig. CS72. Turn oil pump drive shaft until flat on drive gear coupling flange is toward camshaft as shown in Fig. CS73 and install oil pump. Now observe upper end of oil pump coupling as shown in Fig. CS74 where it will be noted that the distributor drive slot is parallel to crankshaft and that slot is offset toward outside of engine. Time the ignition distributor as outlined in paragraph 201A.

OIL PRESSURE RELIEF VALVE

Model 40D4

81. The oil pressure relief valve ball (65—Fig. CS69) is located on delivery side of oil pump body. The spring

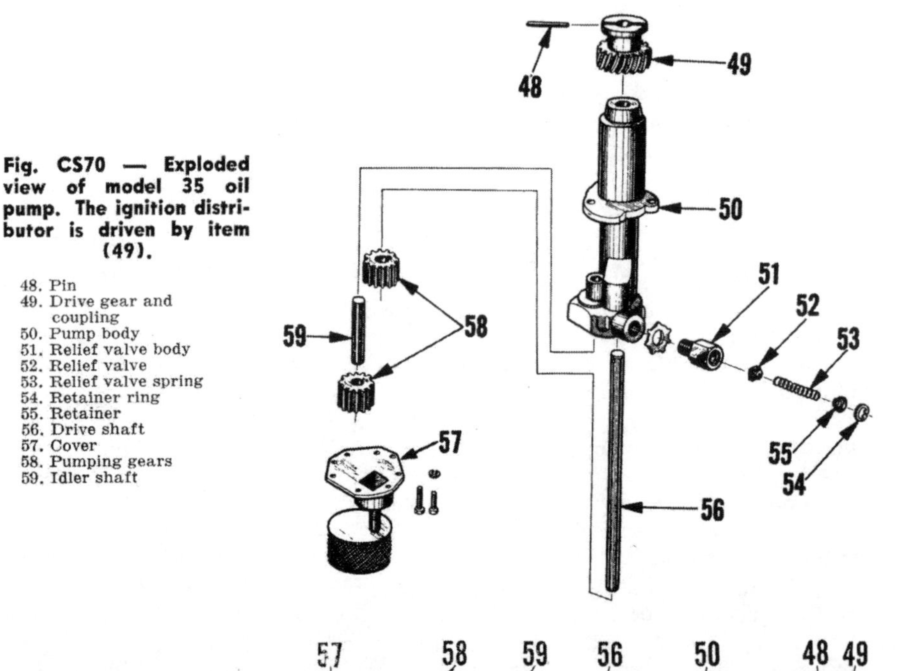

Fig. CS70 — Exploded view of model 35 oil pump. The ignition distributor is driven by item (49).

48. Pin
49. Drive gear and coupling
50. Pump body
51. Relief valve body
52. Relief valve
53. Relief valve spring
54. Retainer ring
55. Retainer
56. Drive shaft
57. Cover
58. Pumping gears
59. Idler shaft

Fig. CS69—Exploded view of model 40D4 oil pump, associated tubing and connections.

53. Driver gear
54. Pump body
55. Bushings
56. Oil pump drive shaft
57. Pumping gears
58. Idler gear shaft
59. Dowel
60. Housing
61. Dowel
62. Gasket
63. Cover
64. Valve body
65. Relief valve ball
66. Relief valve spring
67. Relief valve plug
68. Cotter pin
69. Suction pipe, late models
70. Suction pipe, early models

Fig. CS71 — Sectional view of model 35 oil pump showing the installation dimension for the pump driving gear (49). Refer to legend under Fig. CS70.

loaded valve is designed to maintain a normal oil pressure of 25-35 psi with oil warm and engine running at 1850 rpm. If the pressure is not as specified, remove the oil pan as outlined in paragraph 76A, remove cotter pin (68) and turn plug (67) in to increase or out to decrease oil pressure.

Model 35

81A. As shown in Fig. CS70, the oil pressure relief valve (52) is mounted in the oil pump body. With the oil at normal operating temperature, the relief valve is calibrated to maintain an operating pressure of 30-45 psi. If pressure is not as specified, remove the oil pan, thoroughly clean the valve and add washers to increase the relief valve spring pressure.

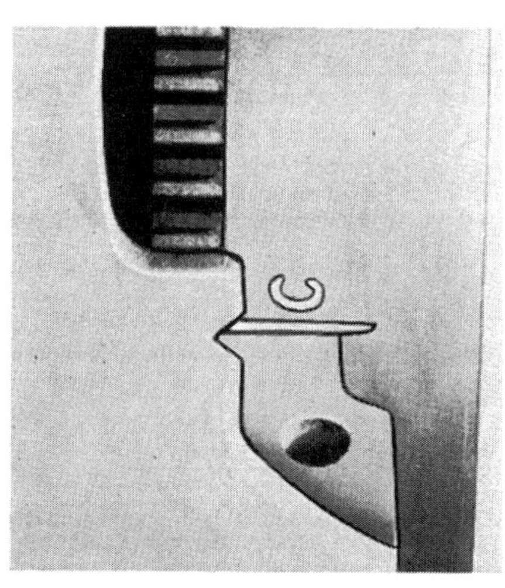

Fig. CS72—Model 35 Flywheel mark "DC" in line with notch in flywheel housing.

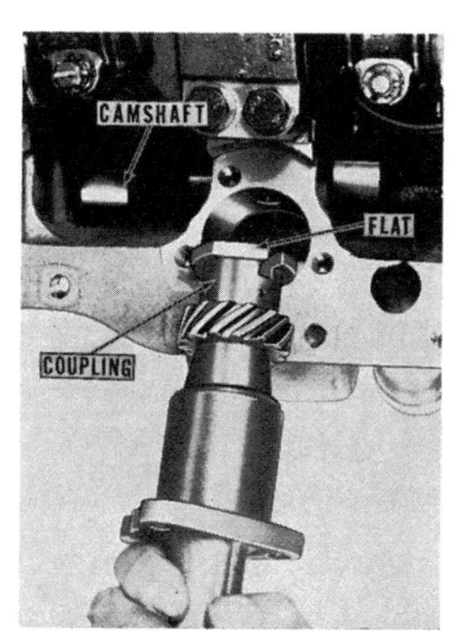

Fig. CS73 — When installing model 35 oil pump, make certain that flat on drive coupling flange is toward camshaft.

CARBURETOR

Model 35

100A. Zenith carburetor model 267J8, outline No. 11973 is used. Calibration and adjustment data are as follows:

Float setting .... $1\frac{5}{32}$ inches
Repair Kit K-11973
Gasket Kit C181-325
Inlet Needle & Seat C81-1-35
Idle Jet C55-22-12
Main Jet C52-7-28
Discharge Jet C66-100-70
Well Vent Jet C77-18-22

Fig. CS74 — Top view of oil pump drive gear and coupling. Notice that narrow side of coupling is away from engine.

DIESEL SYSTEM

Model 40D4

The Diesel fuel system consists of three basic components; the fuel filters, injection pump and injection nozzles. When servicing any unit associated with the fuel system, the maintenance of absolute cleanliness is of utmost importance. Of equal importance is the avoidance of nicks or burrs on any of the working parts.

Probably the most important precaution that service personnel can impart to owners of Diesel powered tractors, is to urge them to use an approved fuel that is absolutely clean and free from foreign material. Extra precaution should be taken to make certain that no water enters the fuel storage tanks. This last precaution is based on the fact that all Diesel fuels contain some sulphur. When water is mixed with sulphur, sulphuric acid is formed and the acid will quickly erode the closely fitting parts of the injection pump and nozzles.

131. **QUICK CHECKS—UNITS ON TRACTOR.** If the Diesel engine does not start or does not run properly, and the Diesel fuel system is suspected as the source of trouble, refer to the Diesel System Trouble Shooting Chart and locate points which require further checking. Many of the chart items are self-explanatory; however, if the difficulty points to the fuel filters, injection nozzles and/or injection pump, refer to the appropriate paragraphs which follow:

DIESEL SYSTEM TROUBLE SHOOTING CHART

	Sudden Stopping of Engine	Lack of Power	Engine Hard to Start	Irregular Engine Operation	Engine Smokes or Knocks	Excessive Engine Speeds	Excessive Fuel Consumption
Lack of fuel	★	★	★	★			
Water or dirt in fuel	★	★	★	★			
Clogged fuel lines	★	★	★	★			
Inferior fuel	★	★	★	★			
Faulty transfer pump	★	★	★	★			
Faulty injection pump timing		★	★	★	★		★
Air traps in system	★	★	★	★			
Clogged fuel filters	★	★	★	★			
Deteriorated fuel lines	★						★
Air leak in suction line	★						
Faulty nozzle				★	★		★
Sticking pump plunger		★		★			
Binding pump control rod				★			
Weak or broken governor springs				★			
Fuel delivery valve not seating properly				★			
Weak or broken transfer pump plunger spring		★	★				
Air leak in venturi, vacuum pipe or governor diaphragm				★		★	

FUEL FILTERS

Model 40D4

132A. **CIRCUIT DESCRIPTION AND MAINTENANCE.** Fuel from the fuel tank flows through a glass sediment bowl (SB—Fig. CS152) which should be removed, drained and cleaned each day prior to starting the engine. The fuel then flows to the primary filter (PF). The drain plug (P) at bottom of primary filter should be removed and a small quantity of fuel drained each day prior to starting the engine. Turn the "T" handle on top of filter daily to clean the element. Periodically, remove the primary filter body and thoroughly wash same with clean fuel.

From the primary filter, the fuel passes through the transfer (lift) pump (TP—Fig. CS152A) to the renewable element type secondary filter (SF—Fig. CS152). If any signs of water are apparent when draining fuel from the primary filter, a small quantity of fuel must be drained from the secondary filter. Renew the filter element every 1000 hrs. or more often in severe dust conditions. The fuel then passes through the third or final stage renewable element type filter (FF) and on to the injection pump.

If any signs of water are apparent when draining fuel from the secondary filters, a small quantity of fuel must be drained from the final filter. Renew the filter element every 1000 hrs. or more often in severe dust conditions.

BLEEDING

Model 40D4

132B. When fuel lines have been disconnected, or when the flow of fuel has been interrupted, bleed trapped air from the system as follows:

Loosen plug (A—Fig. CS152) on top of the secondary fuel filter (SF), turn on the fuel and operate the hand priming pump (PP — Fig. CS152A) slowly until the fuel flows, free of air bubbles, from the plug hole; then tighten plug (A). Loosen plug (B—Fig. CS152) on top of final fuel filter (FF) and operate the hand priming pump slowly until the fuel flows, free of air bubbles, from the plug hole; then tighten plug (B). Loosen plug (C—Fig. CS152A) on the injection pump and operate the hand priming pump slowly until the fuel flows, free of air bubbles, from the plug hole; then tighten plug (C). Crank engine several revolutions to prime the pressure side of system.

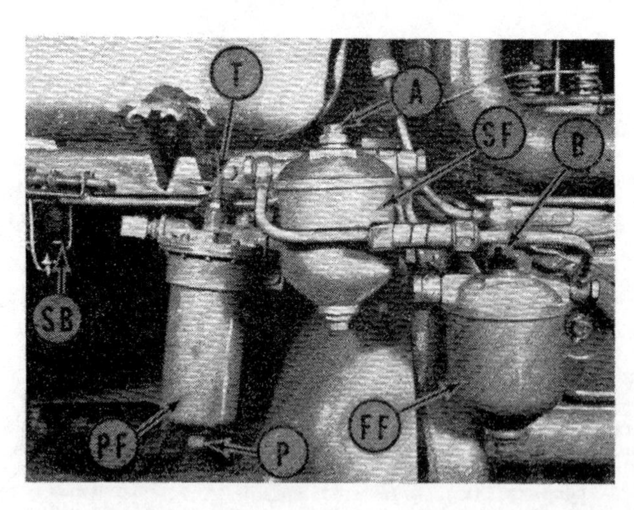

Fig. CS152—Model 40D4 fuel filtering system, showing the primary filter (PF), secondary filter (SF) and final filter (FF).

Fig. CS152A—Left side view of model 40D4 engine, showing the installation of the fuel injection pump and nozzles.

INJECTION NOZZLES

Model 40D4

WARNING: Fuel leaves the injection nozzles with sufficient pressure (1764 psi) to penetrate the skin. When testing, keep your person clear of the nozzle spray.

133A. **TESTING AND LOCATING A FAULTY NOZZLE.** If the engine does not run properly, and a faulty injector is suspected, locate the faulty unit as follows:

If one engine cylinder is misfiring, it is reasonable to suspect a faulty injector. Generally, a faulty injector can be located by loosening the high pressure line fitting on each nozzle holder in turn, thereby allowing fuel to escape at the union rather than enter the cylinder. As in checking spark plugs in a spark ignition engine, the faulty unit is the one which, when its line is loosened, least affects the running of the engine.

134A. Remove the suspected injector from the engine as outlined in paragraph 139A. If a suitable nozzle tester is available, check the unit as in paragraphs 135A, 136A, 137A and 138A. If a tester is not available, reconnect the fuel line to the injector and with the nozzle tip directed where it will do no harm, crank the engine with the starting motor and observe the nozzle spray patterns.

If the two spray patterns are ragged, unduly wet, streaky and/or not symmetrical or, if nozzle dribbles, the nozzle valve is not seating properly. Send the complete nozzle and holder assembly to an official Diesel service station for overhaul.

135A. **NOZZLE TESTER.** A complete job of testing and adjusting the injector requires the use of a special tester such as that shown in Fig. CS153. The nozzle should be tested for opening pressure, seat leakage and spray pattern. Operate the tester lever until oil flows and attach the nozzle and holder assembly.

NOTE: Only clean, approved testing oil should be used in the tester tank.

Close the tester valve and apply a few quick strokes to the lever. If undue pressure is required to operate the lever, the nozzle valve is plugged and same should be serviced as in paragraph 141.

136A. OPENING PRESSURE. While operating the tester handle, observe the gage pressure at which the spray occurs. The gage pressure should be 1764 psi. If the pressure is not as specified, remove the nozzle protecting cap, exposing the pressure adjusting screw and locknut. Loosen the locknut and turn the adjusting screw as shown in Fig. CS153 either way as required to obtain an opening pressure of 1764 psi.

137A. SEAT LEAKAGE. The nozzle valve should not leak at a pressure less than 1600 psi. To check for leakage, actuate the tester handle slowly and as the gage needle approaches 1600 psi, observe the nozzle tip for drops of fuel. If drops of fuel collect at pressures less than 1600 psi, the nozzle valve is not seating properly and same should be serviced as in paragraph 141.

138A. SPRAY PATTERN. Operate the tester handle at approximately 100 strokes per minute and observe the nozzle spray patterns. If the spray patterns are unduly wet, streaky and/or ragged, the nozzle valve should be serviced as in paragraph 141.

139A. **REMOVE AND REINSTALL.** Remove center hood, then before loosening any fuel lines, wash the nozzle holder and connections with clean Diesel fuel or kerosene. After disconnecting the high pressure and leak-off lines, cover open ends of connections with composition caps or tape to prevent the entrance of dirt or other foreign material. Remove the nozzle screws and carefully withdraw the nozzle from cylinder head, being careful not to strike the tip end of the nozzle against any hard surface.

Thoroughly clean the nozzle recess in the cylinder before reinserting the nozzle and holder assembly. It is important that the seating surfaces of recess be free of even the smallest particle of carbon which could cause the unit to be cocked and result in blowby of hot gases. No hard or sharp tools should be used for cleaning. A piece of wood dowel or brass stock properly shaped is very effective. Do not reuse the copper ring gasket located between nozzle and head, always install a new one. Tighten the nozzle holder screws to a torque of 14-16 ft.-lbs.

141. **MINOR OVERHAUL (CLEANING) OF NOZZLE VALVE AND BODY.** Hard or sharp tools, emery cloth, crocus cloth, grinding compounds or abrasives of any kind should NEVER be used in the cleaning of nozzles. A nozzle cleaning and maintenance kit is available through any C. A. V. Service Agency under the number of ET. 141.

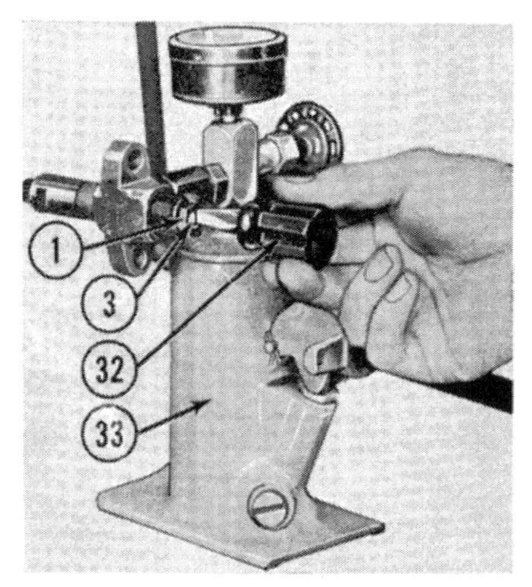

Fig. CS153—Using a suitable tester to check and adjust the nozzle opening pressure.

1. Nut
3. Adjusting screw
32. Screw driver
33. Nozzle tester

Wipe all dirt and loose carbon from the nozzle and holder assembly with a clean, lint free cloth. Carefully clamp nozzle holder assembly in a soft jawed vise and remove the protecting cap (13—Fig. CS153A). Loosen jam nut (30) and back-off the adjusting screw (31) enough to relieve load from spring (5). Remove the nozzle cap nut (7) and nozzle body (9). Normally, the nozzle valve (8) can be easily withdrawn from the nozzle body. If the valve cannot be easily withdrawn, soak the assembly in fuel oil, acetone, carbon tetrachloride or similar carbon solvent to facilitate removal. Be careful not to permit the valve or body to come in contact with any hard surface.

Examine the nozzle body and remove any carbon deposits from exterior surfaces using a brass wire brush (C. A. V. No. ET. 068). The nozzle body must be in good condition and not blued due to overheating.

All polished surfaces should be relatively bright, without scratches or dull patches. Pressure surfaces (A, B and D—Fig. CS153B) must be absolutely clean and free from nicks, scratches or foreign material, as these surfaces must register together to form a high pressure joint.

Clean out the small fuel feed channels (G), using a small diameter wire. Insert the special groove scraper (C. A. V. No. ET. 071) into nozzle body until nose of scraper locates in the fuel gallery. Press nose of scraper hard against side of cavity and rotate scraper to clean all carbon deposits from the gallery. Using seat scraper (C. A. V. No. ET. 070), clean all carbon from valve seat (J) by rotating and pressing on the scraper.

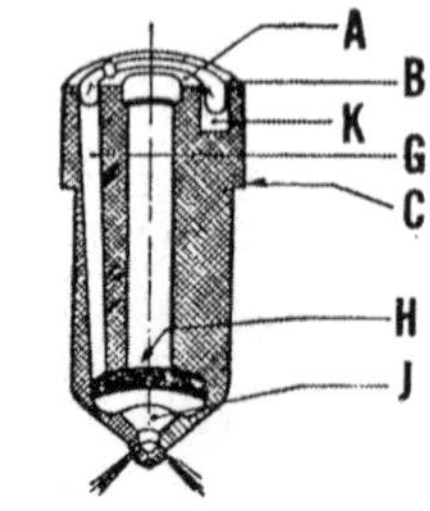
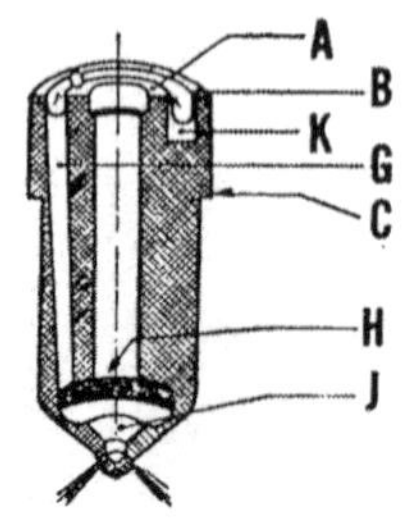

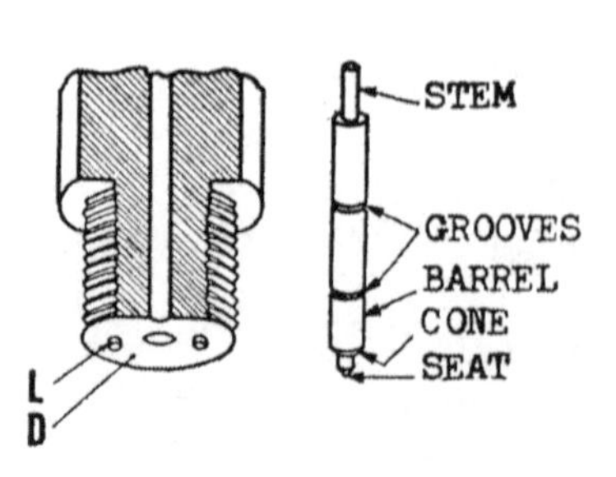

Fig. CS153B — Model 40D4 injection nozzle components, showing points for detailed cleaning and inspection.

Using spray hole cleaner (C. A. V. No. ET. 120) and appropriate size probe, thoroughly clean the two spray holes in the nozzle body end.

Examine the stem and seat end of the nozzle valve and remove any carbon deposits using a clean, lint free cloth. Use extreme care, however, as any burr or small scratch may cause valve leakage or spray pattern distortion. If valve seat has a dull circumferential ring indicating wear or pitting or if valve is blued, the valve and body should be turned over to an official Diesel service station for possible overhaul.

Before reassembling, thoroughly rinse all parts in clean Diesel fuel and make certain that all carbon is removed from the nozzle holder nut. Install nozzle body and holder nut, making certain that the valve stem is located in the hole of the holder body. Tighten the holder nut.

NOTE: Over-tightening may cause distortion and subsequent seizure of the nozzle valve.

Test the injector as in paragraphs 135A, 136A, 137A and 138A. If the nozzle does not leak under 1600 psi, and if the spray patterns are satisfactory, the nozzle is ready for use. If the nozzle will not pass the leakage and spray pattern tests, renew the nozzle valve and seat, which are available only in a matched set; or, send the nozzle and holder assembly to an official Diesel service station for a complete overhaul which includes reseating the nozzle valve cone and seat.

151. **OVERHAUL OF NOZZLE HOLDER.** (Refer to Fig. CS153A). Remove cap (13). Remove jam nut (30) and adjusting screw (31). Withdraw nut (4), spring (5) and spindle (6). Thoroughly wash all parts in clean Diesel fuel and examine the end of the spindle which contacts the nozzle valve stem for any irregularities. If the contact surface is pitted or rough, renew the spindle. Renew any other questionable parts.

Reassemble the nozzle holder and leave the adjusting screw locknut loose until after the nozzle pressure has been adjusted as outlined in paragraph 136A.

Fig. CS153A — Sectional view of a C. A. V. injection nozzle typical of that used on Model 40D4. Nozzle opening pressure should be 1764 psi.

4. Spring cap nut
5. Spring
6. Valve spindle
7. Cap nut
8. Valve
9. Body
10. Inlet connection
11. Leak-off connection
12. Special copper washer
13. Protecting cap
30. Nut
31. Adjusting screw

INJECTION PUMP

Model 40D4

Model 40D4 tractors are equipped with a C. A. V. multiple plunger injection pump.

The subsequent paragraphs will outline ONLY the injection pump service work which can be accomplished without the use of special costly pump testing equipment. If additional service work is required, the pump should be turned over to an official Diesel service station for overhaul. Inexperienced service personnel should never attempt to overhaul a Diesel injection pump.

152A. **TIMING TO ENGINE.** To check and/or adjust the injection pump timing, crank engine until number one piston is coming up on compression stroke and continue cranking until the "DC" line on flywheel is in register with pointer which can be viewed through inspection port in left side of clutch housing cover. Note: The "DC" mark is stamped on rear face of flywheel and can be viewed, using an inspection mirror.

At this time, the "S" marked line on the fuel injection pump gear adapter should be in register with the scribed line on pointer extending from front face of pump as shown in Fig. CS154.

If the "S" line is not in register with the scribed line on pointer, remove the inspection cover from front face of the timing gear case cover and loosen the three cap screws retaining the injection pump drive gear to the gear adapter. Grasp the protruding rear end of the injection pump camshaft with a pair of pliers, turn the shaft slightly to align the "S" marked line on the adapter with the scribed line on the pointer and while holding the pump camshaft in this position, tighten the three cap screws retaining the pump drive gear to the injection pump gear adapter.

Reinstall the inspection cover, making certain that lug on hour meter engages pin on drive plate fastened to the injection pump drive gear.

154A. **TRANSFER PUMP.** Sectional views of the fuel transfer (lift) pump are shown in Fig. CS155. The pump is of the spring-returned plunger type which is driven from the injection pump camshaft.

If the pump is not operating properly, the complete pump can be renewed

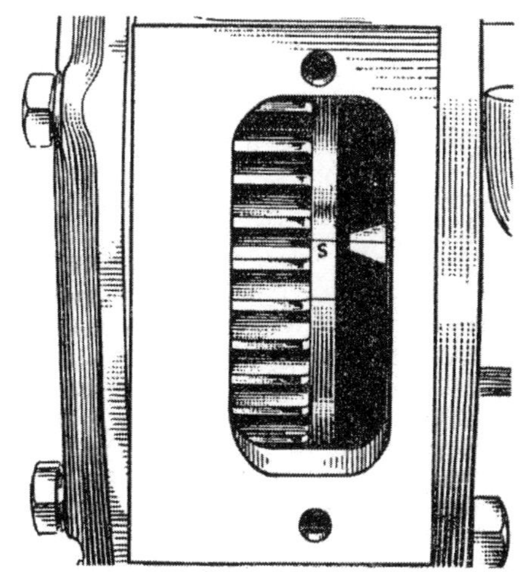

Fig. CS154—Model 40D4. "S" marked line on injection pump drive gear and scribed line on pointer in register.

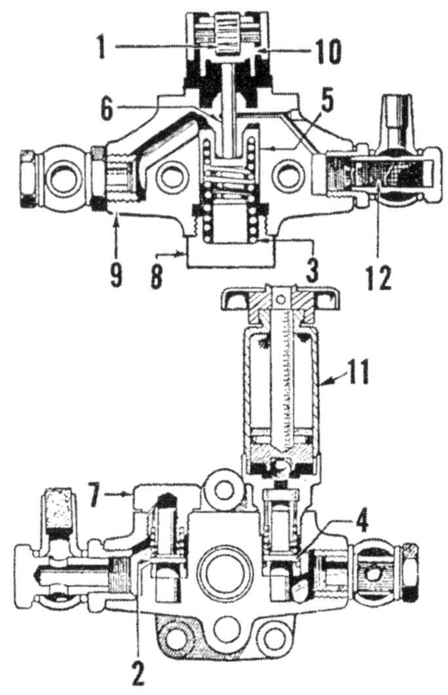

Fig. CS155 — Sectional views of the fuel transfer pump used on model 40D4. Faulty pump operation can sometimes be corrected by a thorough cleaning.

1. Tappet roller
2. Outlet valve
3. Plunger spring
4. Inlet valve
5. Plunger
6. Guide spindle
7. Valve plugs
8. Spring chamber caps
9. Pump body
10. Spindle guide
11. Primer
12. Inlet plug filter gauze

as a unit; or, the transfer pump can be disassembled and cleaned, and checked for improved performance. Quite often, a thorough cleaning job will restore the pump to its original operating efficiency.

156A. **REMOVE AND REINSTALL INJECTION PUMP.** To remove the injection pump, thoroughly wash the pump and connections with clean Diesel fuel. Disconnect the injection pump fuel inlet line, vacuum line and stop cable. Unbolt and remove pump from engine.

When installing the injection pump, first remove the inspection cover from front face of the timing gear case cover and proceed as follows:

Crank engine until number one piston is coming up on compression stroke and continue cranking until "DC" line on flywheel is in register with pointer which can be viewed through inspection port in left side of clutch housing cover. Note: The "DC" mark is stamped on rear face of flywheel and can be viewed, using an inspection mirror. Turn the injection pump drive gear clockwise, viewed from front, until the "S" marked line on the pump gear adapter is in register with the scribed line on pointer extending from front face of pump housing and while holding the drive gear in this position, install pump on engine and tighten the retaining screws.

Note: The pump timing lines may, and very likely will, go out of alignment as the pump gear meshes with its driving idler gear; in which case, adjust the pump timing as outlined in paragraph 152A.

Reconnect the fuel lines, vacuum line and stop cable and when installing the inspection cover, make certain that lug on hour meter engages pin on drive plate fastened to the injection pump drive gear.

159. **GOVERNOR.** The C. A. V. injection pump is equipped with a pneumatic type governor which is actuated by vacuum in a venturi in the engine air induction system.

159A. ADJUSTMENT. Recommended governed speeds are as follows:

Max. no load
engine speed ....1850 rpm
Max. full load
engine speed ....1650 rpm
Belt pulley speed at
1650 engine rpm..........1000 rpm
Power take-off speed at
1650 engine rpm.......... 530 rpm
Low idle engine speed....... 700 rpm

To adjust the governed speeds, first start engine and run until engine is at normal operating temperature, then loosen the lock nut and backout screw (S—Fig. CS152A) several turns. Move the speed control hand lever to wide open position and turn screw (X—Fig. CS156) either way as required to obtain an engine high idle (no-load) speed of 1825 rpm; then turn screw (S—Fig. CS152A) in to increase the high idle speed to 1850 rpm and tighten the lock nut. Move the speed con-

Fig. CS156—Model 40D4 air intake venturi, showing the location of the engine speed adjusting screws.

Fig. CS157 — Removing the governor diaphragm housing and spring from model 40D4 fuel injection pump.

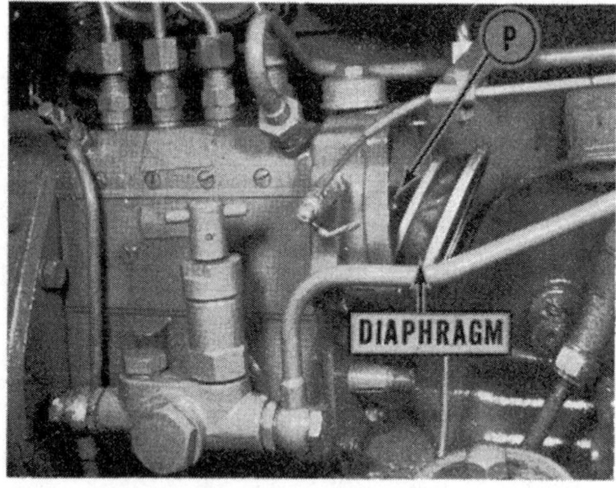

Fig. CS158 — On model 40D4, the governor diaphragm unit can be removed from the injection pump after extracting cotter pin (P).

trol hand lever to the slow speed position and turn screw (Y—Fig. CS156) either way as required to obtain an engine slow idle speed of 700 rpm.

If satisfactory governor operation cannot be obtained, check for a leak in the vacuum line connecting governer housing to venturi or for a ruptured governor diaphragm. Refer to the following paragraph:

159B. GOVERNOR DIAPHRAGM. The smallest pin hole or fracture in the leather diaphragm will yield unsatisfactory governor operation. To test the unit, disconnect the vacuum line, move the stop lever to stop position, place finger over the diaphragm housing union to seal it off from the atmosphere. Release stop lever. The control should then slowly return to the full speed position after a quick initial movement for a fraction of the distance. If the control returns quickly for the complete distance, either the diaphragm housing is not clamped securely to the governor housing or the diaphragm is leaking.

To renew the governor diaphragm, disconnect the vacuum pipe and remove the screws retaining the diaphragm housing to the injection pump. Withdraw the diaphragm housing and spring as shown in Fig. CS157. Using a small pointed instrument, pry the diaphragm retaining rim out of housing as shown in Fig. CS158; then remove cotter pin (P) and renew the diaphragm unit.

When reassembling, be sure to tighten the diaphragm housing retaining screws securely. Check, and adjust if necessary, the governed speeds as outlined in paragraph 159A.

PRE-COMBUSTION CHAMBERS

Model 40D4

175A. **CLEANING.** The necessity for cleaning the pre-combustion chambers is usually indicated by excessive exhaust smoking, or when fuel economy drops. The procedure for removing the caps (CC — Fig. CS152A) and cleaning carbon from the chambers is conventional.

NON-DIESEL GOVERNOR

Model 35 governor is of the centrifugal flyball type mounted on the camshaft timing gear.

ADJUSTMENT

Model 35

182. Before attempting any governor adjustments, first remove any binding or lost motion from the operating linkage. The adjustable ball and socket joint fitted to front end of governor to carburetor rod should be well lubricated and should be adjusted to provide a snug fit without causing friction. Start engine and warm up to normal operating temperature; then stop engine and disconnect the governor to carburetor rod from throttle lever. With the governor lever in the wide open position, adjust the length of the rod to just enter hole in throttle lever when throttle butterfly is in the wide open position.

Start engine and with the speed control hand lever in the retarded position, turn the throttle stop screw either way as required to obtain an engine slow idle speed of 450-500 rpm. Move the hand lever to the wide open

Fig. CS163—Side view of model 35 engine showing governor linkage and speed adjusting screw (SS).

Fig. CS169 — On model 35, use heavy grease to hold the governor flyballs in position when installing the outer race.

Fig. CS169A — Removing the governor ball driver and inner race from model 35 camshaft gear.

position and turn screw (SS—Fig. CS-163) either way as required to obtain the recommended speeds which follow:

Max. no load
  engine speed ....1850 rpm
Max. full load
  engine speed ....1650 rpm
Belt pulley speed at
  1650 engine rpm..........1000 rpm
Power take-off speed at
  1650 engine rpm.......... 530 rpm

R&R AND OVERHAUL

Model 35

193. To remove the governor, first remove the timing gear cover as outlined in paragraph 57B. Withdraw the governor outer ball race (Fig. CS169) and balls. Remove the nut from front end of camshaft and withdraw the ball driver and inner race as shown in Fig. CS169A. To remove the governor shaft from timing gear cover, remove plug (23—Fig. CS169B) from cover, drift out the taper pin retaining inner lever (34) to shaft and withdraw the shaft. Shaft bushings (25) can be pressed from timing gear cover if renewal is required. After new bushings are installed, install the governor shaft and make certain there is absolutely no binding tendency.

When reassembling, install the ball inner race and driver on camshaft gear hub, then install and tighten the gear retaining nut to a torque of 130 ft.-lbs. Check to be sure the inner race rotates freely on the gear hub and that the race has a fore and aft

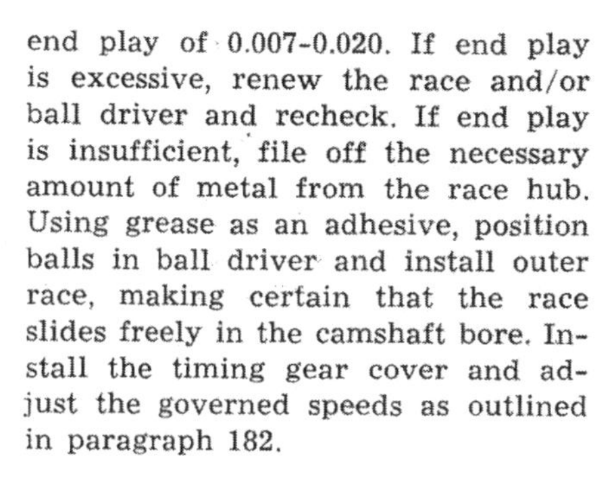

end play of 0.007-0.020. If end play is excessive, renew the race and/or ball driver and recheck. If end play is insufficient, file off the necessary amount of metal from the race hub. Using grease as an adhesive, position balls in ball driver and install outer race, making certain that the race slides freely in the camshaft bore. Install the timing gear cover and adjust the governed speeds as outlined in paragraph 182.

COOLING SYSTEM

RADIATOR

Models 35-40D4

195A. To remove the radiator, first drain cooling system and remove grilles; then remove front and center hood as a unit. Remove adjusting cap from front of steering gear housing and withdraw cotter pin from steering shaft. Loosen the steering shaft center bearing support. Turn the steering worm shaft out of mesh with the sector and pull the shaft forward as far as possible. Then, using a punch, remove the pin retaining steering wheel to steering shaft, remove steering wheel and withdraw the steering worm shaft from tractor. Disconnect radiator hoses, then unbolt and remove radiator.

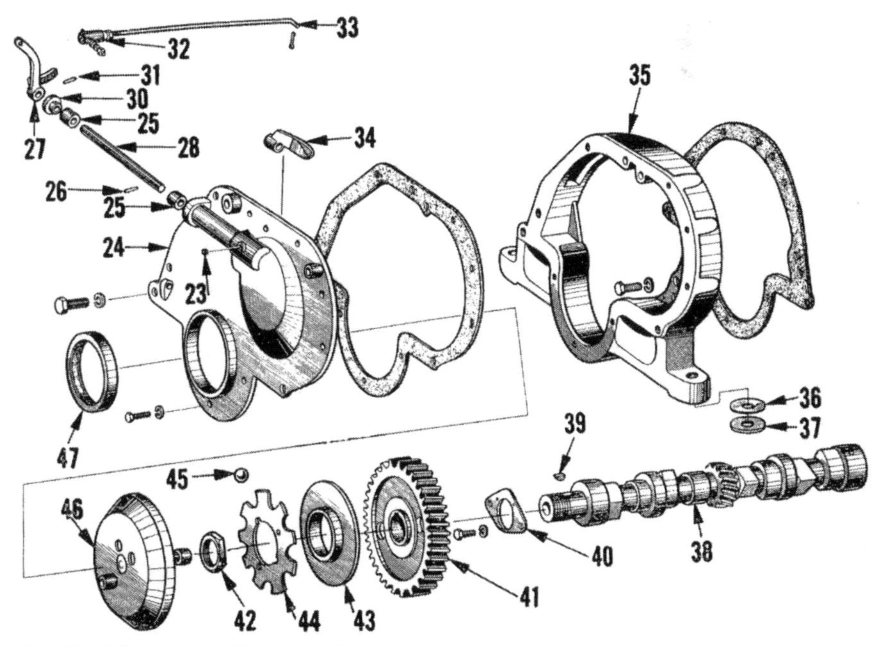

Fig. CS169B—Model 35 camshaft, timing gear case, cover, governor components, etc.

23. Cup plug
24. Timing gear case cover
25. Governor shaft bushings
26. Governor lever pin
27. Outer governor lever
28. Governor shaft
30. Seal
31. Pin
32. Ball joint
33. Governor to carburetor rod
34. Inner governor lever
35. Timing gear case
36. Shim (0.010)
37. Shim (0.005)
38. Camshaft
39. Woodruff key
40. Thrust plate
41. Cam gear
42. Nut
43. Inner race
44. Ball driver
45. Flyball
46. Outer race
47. Oil seal

WATER PUMP

Model 40D4

199A. **R & R AND OVERHAUL.** The procedure for removing the water pump is evident after removing the radiator as outlined in paragraph 195A. To disassemble the pump, remove pulley, rear cover and impeller. Remove snap ring (15—Fig. CS176) and extract rear seal (10 or 22) from pump body. Press the pump shaft out through front of pump body. The remaining disassembly procedure is evident.

Thoroughly clean all parts and renew any which are questionable. When reassembling, press rear bearing (18) onto pump shaft and install spacer (17). Then press front bearing (16) into position. Pack the bearings and the space between the bearings half-full with high melting point grease. Install front seal (20) with front plate and housing (19 & 21) to pump body, then press the assembled shaft into pump body and install snap ring (15). Install thrower (14) with flange toward front and rear seal (10 or 22) with carbon toward rear.

Press the impeller onto pump shaft until there is a clearance of 0.015-0.025 between impeller vanes and pump body as shown in Fig. CS177. Note: Clearance can be checked with a feeler gage and the lower limit of 0.015 is preferred.

Assemble the remaining parts and install pump by reversing the removal procedure.

Model 35

199B. **R & R AND OVERHAUL.** The procedure for removing the water pump is evident after removing the radiator as outlined in paragraph 195A. To disassemble the pump, remove snap ring (9—Fig. CS178), withdraw pulley (8) and remove snap ring (7). Remove rear cover (1), place pump in a suitable press and press the shaft and bearing assembly (6) out of impeller and pump body. Remove seal from pump body. Thoroughly clean and examine all parts for damage or wear. Shaft and bearing are available as an assembled unit only. Reface sealing surface of impeller if it is grooved or otherwise damaged. Renew any other questionable parts.

When reassembling, press new seal in pump body but be sure to press only on the outer flange of seal. Coat sealing surface with grease. Install shaft and bearing unit by pressing only on outer face of bearing, NOT by pressing on end of shaft. Install snap ring (7). Press impeller on pump shaft until rear face of impeller is 0.010 below gasket face of pump body as shown. Support shaft at impeller end, press on the pulley and install snap ring (9). Install gasket and rear cover. Turn pulley several times to be sure pump shaft does not bind.

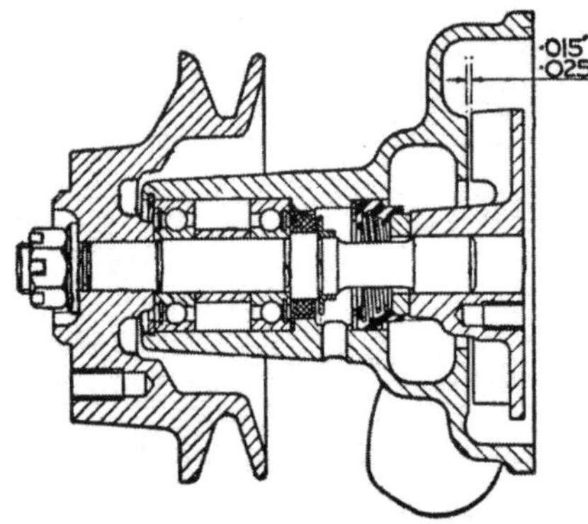

Fig. CS177—Sectional view of water pump typical of that used on model 40D4. When installing the impeller, be sure that there is a clearance of 0.015-0.025 between impeller and pump body.

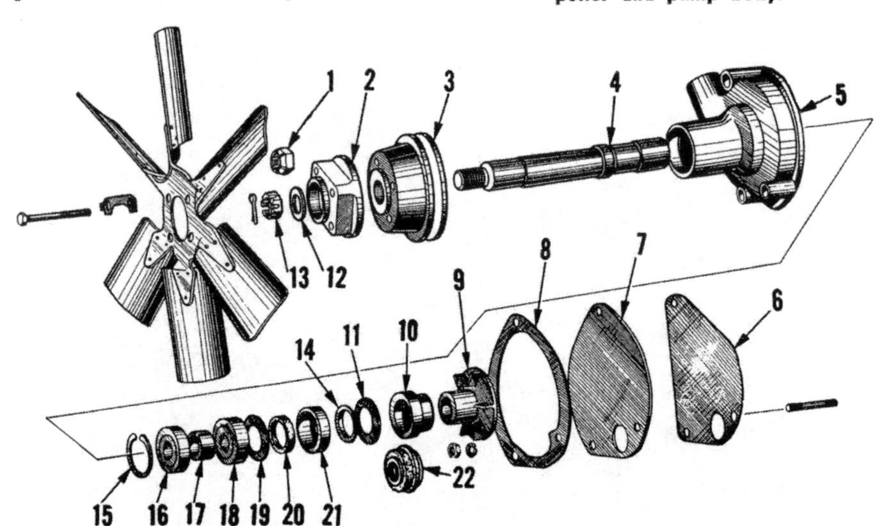

Fig. CS176—Exploded view of model 40D4 water pump showing both the early and late construction.

1. Nut, late models
2. Fan extension
3. Pulley
4. Pump shaft
5. Pump body
6. Gasket
7. Backplate
8. Gasket
9. Impeller
10. Rear seal, early models
11. Rear seal ring
12. Washer
13. Nut, early models
14. Thrower
15. Snap ring
16. Bearing, front
17. Spacer
18. Bearing, rear
19. Front seal plate
20. Felt seal
21. Front seal housing
22. Rear seal, late models

IGNITION AND ELECTRICAL SYSTEM

DISTRIBUTOR

Model 35

200A. Auto-Lite battery ignition distributor No. IAD 4043 is used. Refer to Distributor Section of STANDARD UNITS for general overhaul data. Test specifications are as follows:

Drive end rotation	C.C.
No. cylinders	4
Condenser capacity	.23-.26 mfd.
Cam angle	42°
Advance data (distributor degrees @ rpm)	0° @250
	1° @ 270
	10° @ 475
	14° @ 750
	15° @ 825

201A. **INSTALLATION AND TIMING.** To install the ignition distributor, crank engine until number one piston is coming up on compression stroke and continue cranking until a point on flywheel 2 degrees (1/4-inch) before the "DC" mark is aligned with notch in flywheel housing as shown in Fig. CS184. Examine the oil pump driving coupling and notice that the coupling slot is slightly offset. Turn the distributor drive shaft so that rotor arm is in the number one firing position and mount distributor on engine, making certain that the offset tongue on distributor drive shaft properly engages the offset slot in the oil pump coupling. Adjust the breaker contact gap to 0.020.

Remove cable from No. 1 spark plug and hold free end of cable near engine block. Loosen distributor clamp and, with ignition switch turned on, turn distributor body slowly until a spark occurs at end of spark plug cable; then lock the distributor in this position.

The efficiency and the amount of power that the engine has at rated rpm depends on the proper operation

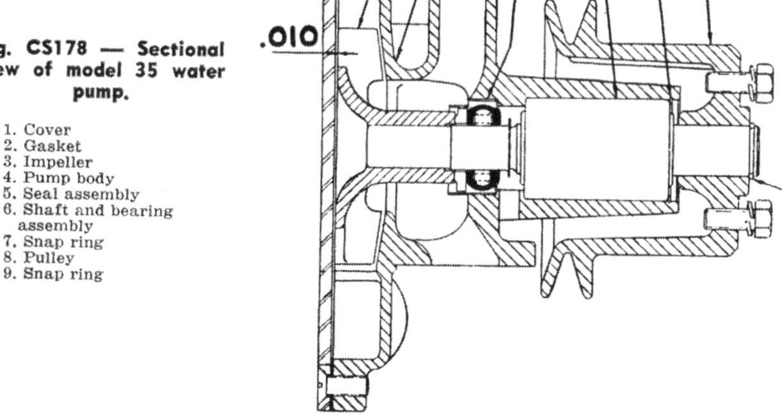

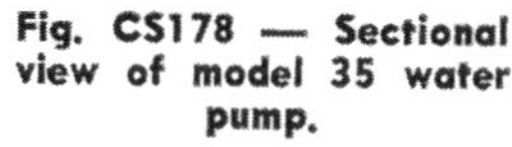

Fig. CS178 — Sectional view of model 35 water pump.

1. Cover
2. Gasket
3. Impeller
4. Pump body
5. Seal assembly
6. Shaft and bearing assembly
7. Snap ring
8. Pulley
9. Snap ring

of the automatic spark advance unit which is built into the distributor. If the engine lacks power at rated rpm, check the distributor advance curve as given in the preceeding specifications.

GENERATOR AND REGULATOR

All Models

202A. Auto-Lite, Delco-Remy and Lucas generators are used and their applications are as follows:

Model 35 ...... Auto-Lite GHD-6001X
Model 40D4, Export prior
Ser. No. 45001 .. Lucas C45P5-22460A
Model 40D4, Export after
Ser. No. 45000 .. Delco-Remy 1100305
Model 40D4, U. S.
and Canadian .. Delco-Remy 1100305

Test specifications are as follows:

Auto-Lite GHD-6001X
Brush spring tension ...... 26-46 oz.
Field draw, volts 5.0
amps 2.0-2.2
Output, volts 8.0
max. amps 19.0

Lucas C45P5
Brush spring tension .... 36-44 oz.
Field draw, volts 12.0
amps 2.0
Output, volts 13.0
max. amps 13.0

Delco-Remy 1100305
Brush spring tension .... 28 oz.
Field draw, volts 12.0
amps 1.58-1.67
Cold output, volts 14.0
amps 20.0
rpm 2300

202B. Auto-Lite, Delco-Remy and Lucas regulators are used and their applications are as follows:

Model 35 ...... Auto-Lite VRR-4103A
Model 40D4, Export prior
Ser. No. 45001 .. Lucas RB-107-37192
Model 40D4, Export after
Ser. No. 45000 .. Delco-Remy 1119191
Model 40D4, U. S.
and Canadian .. Delco-Remy 1119191

Test specifications are as follows:

Auto-Lite VRR-4103A
Ground polarity .... Pos.
Cut-out relay, air gap 0.032
point gap 0.015
closing voltage 6.3-6.8
opening voltage 4.1-4-8
Voltage regulator,
air gap 0.048-0.052
setting volts, range 7.0-7.3
adjust 7.2

Lucas RB-107
Ground polarity .... Pos.
Cut-out relay, air gap .... *
stop gap 0.025-0.030
point gap 0.018
closing voltage 12.7-13.3
opening voltage 8.5-11.0
reverse current ... 3.0-5.0 amps
Regulator, air gap 0.015
setting volts, range 15.3-16.2
adjust 15.7

* To obtain desired air gap, loosen both of the armature retaining screws. Press armature squarely down against the copper-sprayed core face and re-tighten the retaining screws.

Delco-Remy 1119191
Ground polarity .... Pos.
Cut-out relay, air gap 0.020
point gap 0.020
closing volt., range ... 11.8-14.0
adjust 12.8
Voltage regulator, air gap 0.075
setting volts, range 13.6-14.5
adjust 14.0

STARTING MOTOR

All Models

205. Auto-Lite, Delco-Remy and Lucas starting motors are used and their applications are as follows:

Model 35 .... Auto-Lite MCL 6017
Model 40D4, Export prior
Ser. No. 45001 .... Lucas M45J-26112
Model 40D4, Export after
Ser. No. 45000 .. Delco-Remy 1113614
Model 40D4, U. S. and
Canadian prior Ser.
No. 45001 ...... Delco-Remy 1113615
Model 40D4, U. S. and
Canadian after Ser.
No. 45000 ..... Delco-Remy 1113615

Test specifications are as follows:

Auto-Lite MCL-6017
Volts 6
Brush spring tension .... 42-53 oz.
No load test, volts 5.0
max. amps 65
rpm 4900
Lock test, volts 2.0
max. amps 410
torque, ft.-lbs. 8.0

Lucas M45J
Volts 12
Brush spring tension ..... 30-40 oz.
No load test, volts 12
amps 70
rpm 8,000
Lock test, volts 5.2
amps 930
torque, ft.-lbs. 29

Delco-Remy 1113614 & 1113615
Volts 12
Brush spring tension ..... 36-40 oz.
No load test, volts 11.6
amps 95
rpm 8,000
Lock test, volts 2.2
amps 600
torque, ft.-lbs. 20

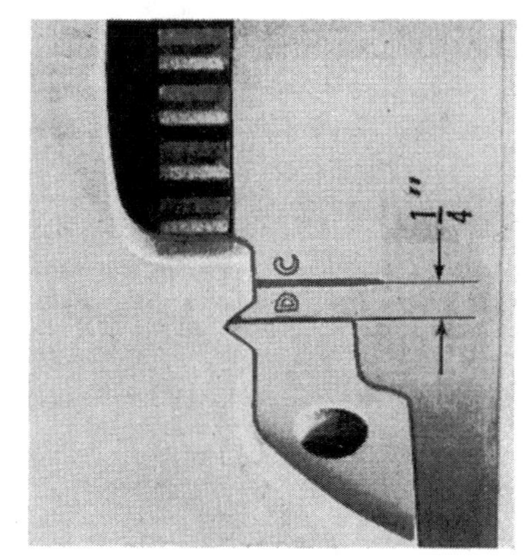

Fig. CS184—Static ignition timing mark for model 35 is 2 degrees (1/4-inch) before "DC".

ENGINE CLUTCH

APPLICATIONS AND OVERHAUL SPECIFICATIONS

Models 35-40D4

210A. Model 35 is equipped with a Borg & Beck model 10A7 clutch which is fitted with a Borg-Warner No. 361323 cover assembly.

Model 40D4 is equipped with a Borg & Beck model 11A6 clutch which is fitted with a Borg-Warner No. 361405 cover assembly.

Refer to Clutch Section of STANDARD UNITS for disassembly, overhaul, reassembly and release lever setting procedures. Pertinent specifications are as follows:

361323 Cover Assembly
- Release lever height.....*$1\frac{7}{8}$ inches
- Pressure spring part no. 305992
  - No. used 3
  - Color ....Brown
  - Pounds test
    - @ height—inches..224-236@$1\frac{11}{16}$
- Pressure spring part no. 4039
  - No. used 6
  - Color ....Light Blue
  - Pounds test
    - @ height—inches..160-170@$1\frac{11}{16}$

*Lever height measured using 11/32-inch (0.340) keystock in lieu of lined plate.

361405 Cover Assembly
- Release lever height....*$1\frac{7}{8}$ inches
- Pressure spring part no. 4039
  - No. used 12
  - Color ....Light Blue
  - Pounds test
    - @ height—inches..160-170@$1\frac{11}{16}$

*Lever height measured using 11/32-inch (0.340) keystock in lieu of lined plate.

ADJUSTMENT

Models 35-40D4

211A. Adjustment to compensate for lining wear is accomplished by adjusting the clutch pedal linkage, NOT by adjusting the position of the clutch release levers. The clutch is properly adjusted when the clutch pedal has a freed travel of 1-$1\frac{1}{8}$ inches. Make this adjustment by varying the length of the clutch release rod (X—Fig. CS202) by means of clevis (Y) at front end of rod.

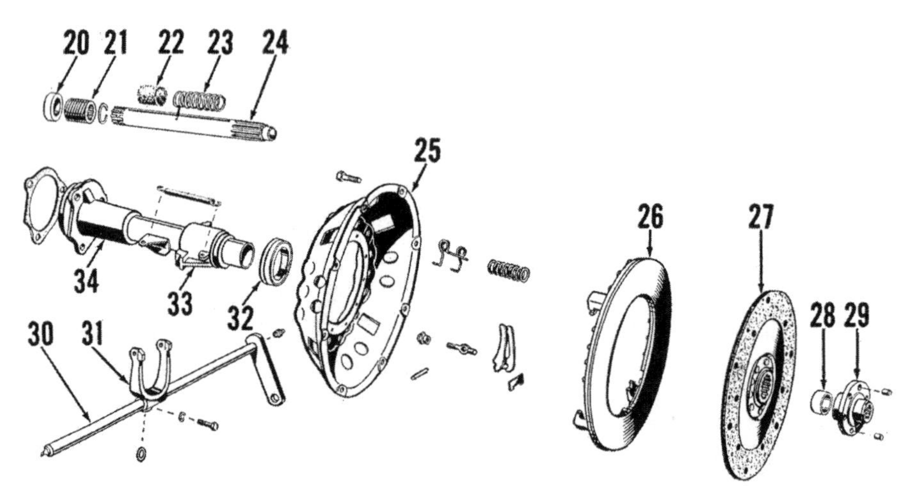

Fig. CS205—Exploded view of model 40D4 clutch, clutch shaft and associated parts. Items (22) and (23) are not used when tractor is equipped with power take-off.

20. Spacer
21. Collar
22. Felt oil seal
23. Spring for seal
24. Clutch shaft
25. Clutch cover plate
26. Pressure plate
27. Lined disc
28. Pilot bearing
29. PTO coupling
30. Release shaft
32. Release bearing
33. Bearing carrier
34. Bearing cap and sleeve

REMOVE AND REINSTALL

Models 35-40D4

218A. To remove the clutch, proceed as follows: Support both halves of tractor separately and disconnect the clutch release rod and the tail light wire. Remove the belt pulley assembly, or remove the pulley hole cover from top of transmission case. Remove the battery, battery box and clutch housing cover. Disconnect the fuel tank rear bracket from transmission housing and block up between fuel tank and frame. Unbolt and split engine frame from transmission housing. Remove the cap screws retaining the clutch cover plate to the flywheel and remove the cover assembly and lined plate.

When reinstalling the clutch, use a spare clutch shaft to align the driven plate splines with respect to the clutch shaft pilot bearing and adjust the clutch pedal linkage as outlined in paragraph 211A.

CLUTCH RELEASE BEARING

Models 35-40D4

219A. Model 40D4 is fitted with one release bearing (32—Fig. CS205); whereas, model 35 is fitted with two bearings as shown in Fig. CS205A. The procedure for renewing the release bearing (or bearings), is evident after splitting the engine frame from the transmission as in paragraph 218A.

CLUTCH SHAFT

Models 35-40D4

222A. To renew the clutch shaft, first split the engine frame from the transmission case as in paragraph 218A. Unbolt the transmision input shaft front bearing cap and sleeve from front of transmission case and withdraw the sleeve and clutch shaft.

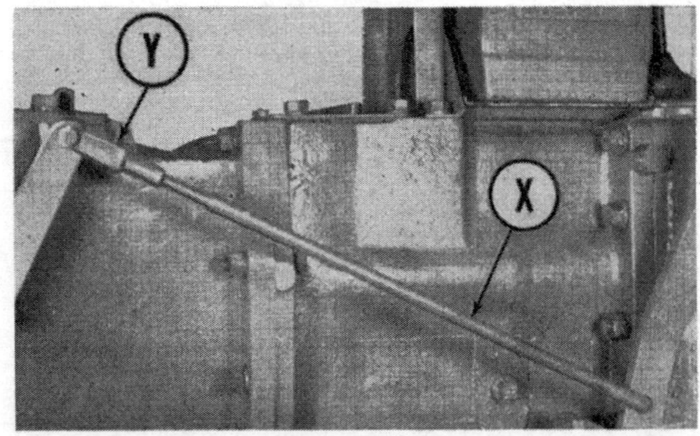

Fig. CS202—Recommended clutch pedal free travel of 1-$1\frac{1}{8}$ inches is obtained by varying the length of rod (X) with clevis (Y).

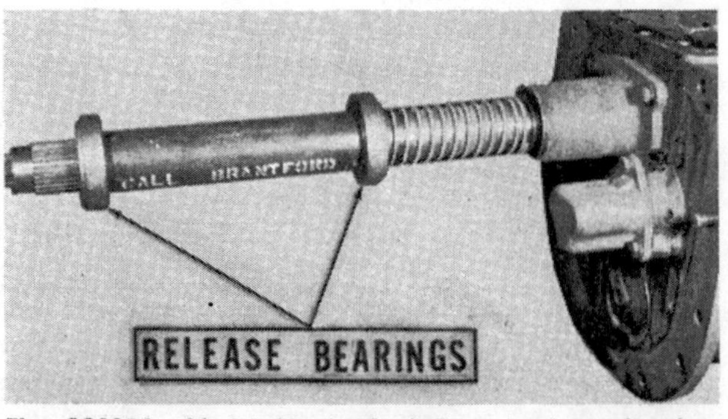

Fig. CS205A—Model 35 clutch shaft and release bearings installations.